Emerging Technologies in Healthcare

This edited book focuses on the role and use of emerging technologies within the healthcare sector. This text draws on the expertise of leading practitioners and researchers who either utilize and/or are at the forefront of researching with emerging technologies in anticipation of enhancing patient outcomes.

Emerging Technologies in Healthcare: Interpersonal and Client-Based Perspectives focuses on the role of emerging technologies in society and how it may enhance medical treatment, management, and rehabilitation of service users. It offers expert perspectives on topics covering emerging technological advances and how they are being incorporated into healthcare, but also critically appraises forthcoming implementation. The editors draw from recent publications and the growing narrative surrounding technological advances, notably telerehabilitation, virtual reality, augmentation, and mHealth. Subsequent chapters focus on these, coupled with other emerging technologies, providing detailed insights into how these can either enhance and/or hinder patient/service user outcomes. Each chapter explores the multifaceted use and application of each emerging technology, that impacts on diagnosis, treatment, and (self-) management of individuals. For example, can emerging technology really facilitate patient diagnosis, improve, or remove practitioner–patient interactions, provide sound rehabilitation, and treat/monitor mental health conditions?

This edited volume encompasses an array of emerging technologies that will remain pertinent to caregivers, families, practitioners, service users, and policymakers. This is not a text on emerging technology alone but on its societal implications, accompanied by ethical, altruistic, and moral examples for such advances within the healthcare field. It is targeted that this text will enhance and offer original discussions surrounding the interconnectivity of technology and medicine, rehabilitation, and patient care.

Rehabilitation Science in Practice

Series Editors:

Marcia J. Scherer
Institute for Matching Person & Technology, Webster, New York, USA

Dave Muller
University of Suffolk, UK

Principles and Practice For more information about this series, please visit:
https://www.crcpress.com/Rehabilitation-Science-in-Practice-Series/book-series/
CRCPRESERIN

Emerging Technologies in Healthcare

Interpersonal and Client-Based Perspectives

Edited by
Christopher M. Hayre, Dave Muller, Marcia Scherer,
Paul M.W. Hackett, and Ava Gordley-Smith

CRC Press
Taylor & Francis Group
Boca Raton London New York

CRC Press is an imprint of the
Taylor & Francis Group, an **informa** business

Cover image: © Shutterstock

First edition published 2024
by CRC Press
2385 NW Executive Center Drive, Suite 320, Boca Raton FL 33431

and by CRC Press
4 Park Square, Milton Park, Abingdon, Oxon, OX14 4RN

CRC Press is an imprint of Taylor & Francis Group, LLC

ISBN: 978-1-032-21578-5 (hbk)
ISBN: 978-1-032-22498-5 (pbk)
ISBN: 978-1-003-27278-6 (ebk)

DOI: 10.1201/9781003272786

Typeset in Times
by Deanta Global Publishing Services, Chennai, India

Christopher Hayre would like to dedicate this book to Charlotte, without whom this would not be possible. He would also like to dedicate this book to his daughters Ayva and Ellena, and in memory of Evelynn. Love to all.

Dave Muller would like to dedicate this to his extended family with love: Pam, Emily, Pete, Lucy, Simon, Tasha, Harlie, Luke, Toby, Edie, Kaya, and Freya.

Marcia Scherer would like to dedicate this to her husband and colleagues from whom she continues to learn so much. Further, she dedicates this to the users and providers of today and tomorrow in their quest to employ technologies wisely and in ways that add quality to life.

Paul Hackett would like to dedicate this book to his wife Jessica.

Ava Gordley-Smith dedicates this to her co-editors and thanks them endlessly for their support, guidance, and enthusiasm.

Collectively, the editors congratulate one another for completing a worthwhile and enjoyable collaborative project.

Contents

Acknowledgments

The editors would like to thank all contributing authors for sharing their innovative work in the field of emerging technology. This is a growing area of interest for practitioners and your commitment to this book reflects the multifaceted use and theory in this contemporary space. On behalf of the editors, it has been a great pleasure working with you all in bringing together this collection of academic work. The work demonstrates not only the existing virtues of emerging technology in practice, but also highlights developmental opportunities within the field. This has been a thoroughly enjoyable project as we have been able to collaborate with colleagues from various parts of the world. Thank you.

Preface

This book presents the collaborative work of experienced researchers worldwide. It brings together practitioners and researchers who are either utilizing or critically evaluating emerging technology in the fields of health and rehabilitation. The chapters presented here discuss the opportunities and challenges of these technologies within several transnational contexts. This book primarily presents arguments and propositions pertinent to health disciplines, but it will naturally present researchers with methodological insight(s) and how these may resonate within other disciplines. The examination of this topic among our experts seeks to provide greater impact to an array of stakeholders internationally.

This book examines the role of emerging technologies with a focus on interpersonal and client-based perspectives. This looks to demonstrate the wide acceptance and impact of emerging technology and how it can enhance the experience of patients and service users. This book, at times, also seeks to critique emerging technologies as we witness its exponential application. In response, then, this book will offer value to undergraduate and postgraduate students, supported with references signposting students to other key material. For practitioners and prospective researchers in health and rehabilitation, this text will naturally provide methodological and empirical insights, helping to uncover areas of novelty in their field.

Dr Christopher M. Hayre
Professor Dave Muller
Professor Marcia Scherer
Professor Paul M.W. Hackett
Ava Gordley-Smith

About the Editors

Dr Christopher M. Hayre is an Associate Professor in Medical Imaging at the University of Canberra, Australia. He has published over 70 refereed papers in the field of medical imaging, medical anthropology, and emerging/disruptive technologies. He is the founding editor of a Book Series with CRC Press, titled: Medical Imaging in Practice.

Professor Dave Muller is currently the Editor of the CRC series, with Professor Marcia Scherer, on Rehabilitation Science in Practice. He was the founding Editor of the journal *Aphasiology* and is currently Editor-in-Chief of the journal *Disability and Rehabilitation*. He has published over 40 refereed papers and has been involved either as series editor, editor, or author of over 50 books. He is a visiting Professor at the University of Suffolk, United Kingdom.

Professor Marcia Scherer is a rehabilitation psychologist and founding President of the Institute for Matching Person & Technology. She is also a Professor of Physical Medicine and Rehabilitation at, University of Rochester Medical Center, USA, where she received both her Ph.D. and MPH degrees. She is a past member of the National Advisory Board on Medical Rehabilitation Research, National Institute of Health, and is Editor of the journal Disability and Rehabilitation: Assistive Technology. She is Co-Editor of the book series for CRC Press, Rehabilitation Science in Practice. Dr Scherer is a Fellow of the American Psychological Association, American Congress of Rehabilitation Medicine, and the Rehabilitation Engineering and Assistive Technology Society of North America (RESNA). Dr Scherer has authored, edited, or co-edited nine book titles and has published over 80 articles in peer-reviewed journals, 50 published proceedings papers, and 30 book chapters on disability and technology. Her research has been cited more than 5000 times by others.

Paul M.W. Hackett is the originator and developer of the Declarative Mapping Approach to social research and the Declarative Mapping Sentence (DMS) for guiding and interpreting qualitative and philosophical research enquiries. In his research and publications, he considers the use of categorial ontologies/mereologies in the understanding of behavior and experience. Most recently, he has focused on the perception and understanding of abstract fine art and on avian behavior and cognition. In *Facet Theory and the Mapping Sentence: Evolving Philosophy, Use, and Application*, he looks at the development of facet theory, viewing philosophical/psychological categorization using a mapping sentence. He holds professorial appointments at the University of Suffolk, UK and Nnamdi Azikiwe University, Nigeria, and teaches and supervises at the University of Wales, Trinity St David, UK and Emerson College, USA. He has approximately 300 publications, which include 30 books. He holds PhD degrees in psychology and fine art.

Ava Gordley-Smith is conducting her PhD research at the University of Wales Trinity Saint David. Ava holds a visiting scholar position at Nnamdi Azikiwe University in Nigeria, is a member of the British Psychological Society, the Royal Philosophical Society, the Society of Applied Philosophy, is the social media editor of *Disability and Rehabilitation* a Q1/2 Journal and is a student fellow of the Royal Anthropological Institute.

1 Emerging Technologies in Healthcare
Interpersonal and Client-Based Perspectives

Clare Killingback and John Naylor

HEALTH AS WHOLENESS

We shall later go on to define our understanding of the terms "technology" and "person-centered practice" in the context of healthcare, but will begin by discussing what we mean by health. The word "health" has its etymological roots in the old English word "hælth" and closely relates to the "whole" as "a thing that is complete in itself" (Brüssow, 2013). However, this historical view, that to be healthy is to be whole, does not necessarily encompass the "complete wellbeing" aspects of the World Health Organization (1948) definition as it alludes instead to a person as a whole entity. Yet in the context of healthcare, medicine and medical research has tended to be disease-focused rather than holistic. Health in the modern world has been particularly focused on a biomedical approach which has its origins in the seventeenth century.

THE BODY AND MIND DIVIDED

The seventeenth-century philosopher Rene Descartes played a formative role in the development of what we now consider modern medicine. Descartes posited that the body and mind exist as separate entities. In essence, the mind is the core of the human being, and the body is a machine-like vessel for the mind (Keller, 2020, Magee, 1987). This dualistic view has had a profound impact on healthcare which often sees the body as a machine in need of repair (Keller, 2020). This metaphor bears heavily upon the question of what is meant by health and healing (Berry, 1994) and creates a division (the antithesis of wholeness or health) by falsifying the process of healing and the nature of the one needing to be healed (Berry, 1994).

HEALING THE BODY/MIND DIVIDE

Twentieth-century phenomenologist Maurice Merleau-Ponty countered Descartes' reductionist view of the body by presenting a more holistic view of human beings

DOI: 10.1201/9781003272786-1

1

(Keller, 2020). His view of the lived body, where the mind and body are reintegrated as one, is in opposition to that presented by Descartes. Importantly, Merleau-Ponty's vision of the integrated lived body has implications for healthcare (Leder, 1984, Keller, 2020) as we move away from the metaphorical body as a broken machine to be "fixed" to a focus on caring for a human being. Illness and disease here are viewed from a more existential perspective as involving suffering, fear, loss, hope, and change (Keller, 2020).

MODERN HEALTHCARE PRACTICE

Sadly, this holistic view of the human being "as a whole" has not always broken through into modern healthcare practice. Healthcare services promote activities focused on treating a physical illness through assessment, diagnosis, pathology, specialized outcomes, technology, audits, and treatment pathways to promote efficiency, with emotional and psychosocial needs often treated separately, if at all (Keller, 2020). This focus on the physical has been particularly evident in the earlier careers and formal training for the authors' own context of physiotherapy practice. Traditionally physiotherapy has been aligned with biomedical models of practice underpinned by positivist paradigms (Wiles and Barnard, 2001). The adoption of this biomedical or biomechanical view of the body was intentional and important in establishing the legitimacy of the profession in aligning them with medical practitioners to gain public trust (Nicholls and Gibson, 2010). The introduction of the biopsychosocial model in the 1970s (Engel, 1978) challenged the biomedical discourse of reductionism by offering a more holistic alternative (Borrell-Carrió et al., 2004). The physiotherapy profession adopted this paradigmatic shift toward a biopsychosocial model of care which considers the patient as a whole person, including their social, cultural, and environmental context (Sanders et al., 2013).

In more recent times, there has been a further shift in international healthcare conversations, including physiotherapy, to focus more explicitly on person-centered practice and to prioritize this as the core model for care delivery (Groves, 2010, Foot et al., 2014, NICE, 2017, WHO, 2015, van Dulmen et al., 2015). Being person-centered refers to a philosophy in which the values, preferences, and individual perspectives of the person play a central role in how their needs are met with a view to optimizing the experience of care (Jesus et al., 2016). Key principles of person-centered practice include respect; choice and empowerment; patient involvement in health policy; access and support; and information (Groves, 2010). These principles are important because person-centered practice focuses on the whole life requirements, thereby determining what makes life meaningful for an individual (Håkansson Eklund et al., 2019). Person-centered practice is related to health as wholeness rather than reducing it to being predominantly biomedical and physical.

The shift in healthcare paradigms from biomedical, to biopsychosocial, to person-centered is theoretically sound. Yet the reality is that healthcare remains reductive and is not necessarily about the "wholeness" we previously discussed.

Physiotherapy practice and education remain typically entrenched in a biomedical discourse (Mudge et al., 2014, Roskell, 2013, Nicholls and Gibson, 2010, Foster and Delitto, 2011, Brun-Cottan et al., 2020). Narrow outcomes are often prioritized over more meaningful human processes and issues of well-being (Todres et al., 2009). The way that healthcare is organized and practiced means that patients are reporting that they do not feel fully met as human persons and that these human dimensions are important to them (Todres et al., 2009).

A recent position paper from the International Experience Exchange for Patient Organisations ((IEEPO), 2021), a patient-led initiative which aims to improve healthcare around the world, is calling for a transformational change, with humanizing healthcare. They consider humanizing healthcare to be a means to rebuild healthcare systems around the needs of patients and their communities, where people are at the heart of healthcare. Humanized care is defined as working with patients and all stakeholders to create a more personalized approach, as opposed to the impersonal, automated, mechanical procedures which can result from a reliance on technology (IEEPO, 2021, Busch et al., 2019). The IEEPO calls on healthcare professionals to "see" and treat the patient, and not just the illness. And so, we return to the beginning where our understanding of what we mean by health determines how healthcare is delivered and perceived by those receiving healthcare services.

The reader may be wondering how these various philosophical perspectives relate to a chapter on emerging technology in healthcare. Understanding our philosophical beliefs about what health is or how we view patients is central to the role that technology can play in our healthcare practice.

We do not wish to deny the achievements of modern healthcare, including advances around healthcare technology, and must acknowledge the role that this technology has played in improvements to our health and well-being. But we must consider voices, such as Charon, when they speak about the:

> vexing failures of medicine – with its relentless positivism, its damaging reductionism, its appeal to the sciences and not to the humanities in the academy, and its wholesale refusal to take into account the human dimensions of illness and healing
>
> **(Charon, 2006), p. 193**

Health and illness are often assessed in economic terms in relation to policy documents and statistical data from clinical outcomes. But where is the person in all of this? So, it could be fair to question whether something is missing from modern healthcare, despite the technological advances. In contrast to the high-technology world, what is missing might be as simple as seeing patients as people – as whole people and not simply a physical body which is diseased.

TECHNOLOGY IN HEALTHCARE

Next, we must consider what is meant by technology in healthcare. Technology in healthcare is composed of five major categories (Geisler and Heller, 2012), p. 3:

1. Medical devices: equipment, instruments, machines, and other devices used for clinical diagnostics, critical care, and other medical administrative functions (e.g., MRI, X-ray machines).
2. Drugs/pharmaceuticals: compounds used in clinical care, both in the prescription and over-the-counter categories.
3. Disposables: the one-time usage materials and devices which are discarded after use and do not constitute equipment in the medical devices category (e.g., catheters, disposable syringes).
4. Medical/surgical procedures and services: the medical and surgical knowledge involved in carrying out medical/surgical interventions.
5. Information technology: the informatics, automation and computer usage classes or equipment, software and techniques utilized in the clinical and administrative areas of the healthcare environment.

Questions about the role of drugs, pharmaceuticals, or Big Pharma are beyond the scope of this chapter. We also recognize that there is a difference between genetic engineering as a medical intervention and the use of a Zimmer frame as a medical device to support independent mobility. But let us look more closely at two categories of healthcare technology in the form of medical devices and information technology which are closely linked: the Internet of Medical Things and artificial intelligence.

The Internet of Medical Things (IoMT) is a way of connecting medical technology, such as smart devices monitoring heart rate, blood oxygen saturation, or skin temperature, for example, with information technology systems through networking technologies (Singh et al., 2020). The IoMT has the potential to monitor biomedical signals and diagnose disease without human intervention (Vishnu et al., 2020). Thus, there is a network of medical devices and people that utilizes wireless communication to facilitate the exchange of healthcare data (Al-Turjman et al., 2020). For example, wireless sensors can remotely monitor a person's health and draw on communication technologies, such as the internet, to send this information to healthcare workers (Al-Turjman et al., 2020). Linked closely to this is the role that artificial intelligence plays in enabling the IoMT to support healthcare professionals with their clinical decision-making (Al-Turjman et al., 2020). For example, computers can use the data generated by feedback from healthcare staff and patients to learn decision-making, distinguishing normal and abnormal results (Al-Turjman et al., 2020).

The IoMT is not without its challenges. Issues of precision and accuracy of the data obtained by the sensors is vital as inaccurate data could be misleading. A secure system that ensures privacy is maintained is vital to avoid hackers stealing data which would result in identity theft or access to controlled substances. The electrical safety of devices must be properly maintained. The need for wireless devices to operate 24 hours a day raises issues of energy consumption and energy use efficacy. Usability of devices needs to be considered, and data storage solutions need to be realized for the ever-increasing volumes of data generated (Al-Turjman et al., 2020).

Furthermore, there are undoubtedly economic challenges to the continuous technological advances in healthcare. In the twenty-first century, we are more dependent than ever on technology. Some new technologies can result in lower short-term spending; vaccines are one example of this, but, more generally speaking, technological advances in healthcare tend to result in increased spending (Goyen and Debatin, 2009). How do we continue to pay for the cures that get ever more expensive? And how do we ensure that advances in specialization, technology, and drugs do not just become a way of marketing industrial products? Where are people in these conversations?

The challenges listed above mostly include the economic and practical aspects of managing such a system but there are wider ethical aspects to consider. The IoMT indeed needs to be mindful of being technologically and scientifically robust but also ethically responsible and respective of service users' rights (Mittelstadt, 2017). Social isolation could be a risk when healthcare visits from healthcare staff to patients are made redundant because they have been replaced by technological monitoring (Mittelstadt, 2017). Healthcare via remote monitoring is at risk of removing the human element of care which can only be developed via face-to-face relationship. Being lonely is a significant risk factor for premature mortality and has been compared to the effects of smoking (Holt-Lunstad et al., 2010). This raises ethical questions. What is the use of healthcare technology if we are made socially poorer by the result of it? What if our health and well-being were reduced to a narrow range of measurable and quantifiable outcomes that are taken out of the context of being a whole person without consideration of social and mental aspects of our health and well-being? The IoMT may result in greater efficiency, but it may come at the cost of the qualitative aspects of care which are what help to add meaning to our lives. What are the unintended or unanticipated consequences of healthcare surveillance technology? Answering these questions is beyond the scope of this chapter but it is important that these questions are considered.

QUESTIONING ADVANCING TECHNOLOGY

Technology is both ubiquitous and invisible (Haraway, 2000). It has become an inextricable part of today's healthcare practices (Jacobs et al., 2017). We cannot argue with the remarkable progress of health technology in recent decades and the impact this has had in increasing life expectancy. Indeed, the IoMT and artificial intelligence have a role to play in this; the point here is not about being anti-technology, but we do need to consider unpacking some of the broader ethical aspects of technology in today's healthcare practice.

The IoMT and artificial intelligence come from the scientific method of healthcare, but we must consider alternative viewpoints. To do this, we need to look backward to some of the voices of the twentieth century, who were warning about the moral questions of ever-advancing technology. Let us look more closely at the phenomenological viewpoint, as this can offer a different perspective.

Phenomenology seeks to address philosophical questions, which stem from what phenomenologists call the "lived experience." Phenomenology does not begin in the world of science, but in embodied meaning experiences known as the "life world." As humans, we do not only have a body, but everybody is a body, a body which is lived in (Svenaeus, 2018).

Martin Heidegger, the twentieth-century phenomenologist wrote about the risk of technology dominating, being taken for granted, and becoming part of our world view to the point of being barely visible (Svenaeus, 2018). The role of modern technology in society has come to the point where there is no other way to live; again, we must not underestimate the positive role that technology and science have played in healthcare practice. Thus, Heidegger is not saying in his discourse on modern technoscience (Heidegger, 1977) that healthcare professionals need to be averse to science and subsequent technology, but rather they should be aware of the limitations of science and technology and be mindful of the "dangers of acting only as scientists in their profession when they are meeting patients" (Svenaeus, 2018, p. 82). Bioethicists are drawing on the work of phenomenologists to examine the ethical challenges that healthcare technology brings, such that "our abilities to handle new technologies – and not let technologies handle us – will be decisive for the society to come" (Svenaeus, 2018, p. 90).

To say this another way and through another phenomenologist, Gadamer (1996), when discussing medical science, warns of the:

> dissolution of personhood when the patient is objectified in terms of a mere multiplicity of data. In a clinical investigation all the information about a person is treated as if it could be adequately collated on a card index. If this is done in a correct way, then the data all belong to the person. But the question is nevertheless whether the unique value of the individual is properly recognized in this process.
>
> **(Gadamer, 1996) (p. 94)**

The concern of phenomenologists here is a warning for healthcare professionals to ensure that we do not frame human beings through medical science and technology alone, but that we also approach our patient encounters through an embodied approach and human dialog. We must remember to see our patients as people first and foremost and keep technology and medical science in their right place. Otherwise, healthcare technology may pose the risk of "dissolving" the person (Svenaeus, 2018).

This brings us back to Descartes and challenges us as healthcare professionals as to how we see our patients. Do we view them as a scientific biological body or as a lived body? Our answer to this question will determine the type of healthcare professional we will become, and phenomenologists would argue that healthcare needs to acknowledge the priority of the lived body, as one which is embodied and not merely a diseased body. Again, this is not to be hostile towards healthcare technology, but to hold the perspective of the lived body as a "way of being-in-the-world [that] will also make us wary of the technologies that tend to block life-world concerns in order to prolong or even produce life as a goal in itself" (Svenaeus, 2018, p. 86). Therefore we must return to the view that health is related to wholeness; we are more than a physical body, we are a lived body. In essence, it raises an important question: Is it possible to have person-centered healthcare technology with a focus on people and community, as well as being mindful of efficiency?

PERSON-CENTERED TECHNOLOGY

The modern hospital environment is a place that we associate with high technology, alongside related themes of specialization, standardization, protocols, and efficiency. Efficiency, in particular here, relies heavily on the vital functions of unseen and heavily automated technological "machinery." Faced with such technological advancements, we need a reminder that these places serve real people with non-standardized lives, along with their embodied, contextualized, and narrated experiences (Svenaeus, 2018). Consequently, in healthcare settings, the world of the unique must interact with that of the "standard," whatever that means.

Straddling these two worlds, however, are the healthcare professionals: highly trained individuals who aptly draw on their specialized, high-technology knowledge to meet lay people, most often lacking in this knowledge. Thus, when performing at their person-centered best, clinicians might constitute the palatable human interface with this technology. When underachieving in this role, clinicians might unknowingly be conducting themselves, in the mind of the patient at least, as being at one with the machines, in the fashion of the unthinking, unfeeling automatons of science fiction. Healthcare professionals as humans must therefore bridge this chasm.

The paradigms of person-centered practice and healthcare technology could be said to be antithetical. However, instead of focusing on such polarization, clinicians would be better advised to find ways in which to apply the principles of person-centered practice *to* healthcare technology. In this way, person-centered healthcare's focus on the values, preferences, respect, choice, empowerment, involvement, and individual perspectives of the person, in considering how their healthcare needs are met, could be married with technology. We would suggest an urgent need to consider person-centeredness *and* technology together with the understanding that, although the modern healthcare system is technologically shaped, it must be considered through a human lens – with people at the heart of it. This liminal space is where technology *and* person-centered practice can co-exist, preferably in a way that facilitates and thrives rather than hinders (Lapum et al., 2012). Rather than be governed by technology in our healthcare practice, we perhaps need to move toward consciously integrating person-centered practice and recognizing our humanness in a highly technological healthcare environment.

Technology and "humanness" are relational and thus responsible for shaping our way of being as healthcare practitioners (Lapum et al., 2012). At present, technology appears to be the dominant form of knowing in many healthcare services. The authors are therefore keen to champion those healthcare professionals who can harness the benefits of technology but in a manner where people remain at the heart of their practice. There will be times when the emphasis is rightly placed on technology, such as in critical or intensive care scenarios – or simply when this is highly valued by patients, which means flexibility is more important than dogmatic application of our vision. Furthermore, some have cautioned the substitution of Cartesian mind-body dualism with that of technology/people or objective/subjective dichotomies (Lapum et al., 2012). At the risk of being over-simplistic, the authors would argue that there is a space where both person-centered practice and technology can intersect, offering the best in person-centered healthcare technology practice (Figure 1.1).

The metaphor of *travel* between person-centered and healthcare technological paradigmatic worlds, with particular importance placed on the direction of travel, refutes the idea that people cannot cross between them. A lay person's attempt to cross into the professional world would likely be too steep a climb for most, thus necessitating movement by the healthcare professionals into the lay world. The meeting needs to be between two *people,* where healthcare professionals can act as a metaphorical bridge between the professional world and the lay person's world (Figure 1.2).

Sometimes as healthcare professionals, we think that we are crossing that boundary when we provide people with information in lay terms to help them make an informed decision. Atul Gwande (2015) says, when discussing the role of the healthcare professional in end-of-life care, that the mistakes clinicians make is that they see their role as a task in supplying cognitive information, the cold facts, and descriptions in lay terms. However, he says more often people are looking for the meaning behind the information rather than the facts: what might be a person's biggest fears or concerns? What goals are most important?

> Our most cruel failure in how we treat the sick and the aged is the failure to recognise that they have priorities beyond merely being safe and living longer; that the change to shape ones' story is essential to sustaining meaning in life.
>
> **(Gwande, 2015, p. 243)**

These are the wider existential, phenomenological aspects that matter to people's lives beyond the physical body, although we must also be mindful of the physical aspects.

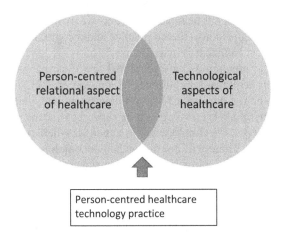

FIGURE 1.1   Person-centered healthcare technology practice.

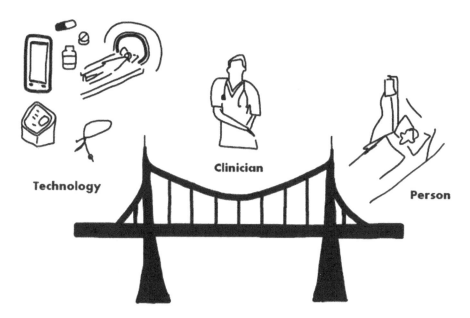

Technology   **Clinician**

                                                              Person

FIGURE 1.2   Clinicians bridging the world of the person and healthcare technology.

As clinicians, we need to be ready to cross the boundary to the world of the person. We know, as healthcare professionals ourselves, that when our professional relationship with our patients has been a meeting between two human beings, these moments have been some of the most rewarding and moving aspects to our work; the things that have made our work meaningful and why we signed up for a caring profession.

ROLE OF HEALTHCARE PROFESSIONALS IN A HIGH-TECHNOLOGY HEALTHCARE SERVICE

Earlier in this chapter, we included a conceptualization of five aspects of healthcare technology (medical devices, drugs/pharmaceuticals, disposables, medical/ surgical procedures and service, information technology). This conceptualization of healthcare technology focuses on material objects. Although this is true, we also need to consider the socially embedded aspects of technology which shape practice (Lapum et al., 2012). For example, healthcare practitioners respond behaviorally and cognitively to interpreting technological readings, resulting in protocols and pathways which shape healthcare practice (Lapum et al., 2010, 2012, O'Keefe-McCarthy, 2009). The thoughtless paraphrasing of a spinal imaging report to a patient in a way that paints the picture of a crumbing spine can create a powerful nocebic effect for the recipient. This emphasizes the important point that clinicians realize the technical words they use without due consideration can cause considerable harm. This implies that it is not just the technology (the object) but our response to the object that matters, and this can potentially be a reductivist, positivist response or a person-centered response.

In many ways, to be human is to be limited – to live within the constraints of our biological bodies and the limits set by the cells which make up our physical body (Gwande, 2015). Technology in healthcare has enabled us to extend some of these physical limitations in a powerful way. Gwande (2015) is calling for a challenge in how we view the role of healthcare professions when he says:

> We've been wrong about what our job is in medicine. We think our job is to ensure health and survival. But really it is larger than that. It is to enable wellbeing. And well-being is about the reason one wishes to be alive. Those reasons matter not just at the end of life, or when debility comes, but all along the way.
>
> **(Gwande, 2015, p. 259).**

This view is important because it calls on healthcare professionals to be cognisant of the power of healthcare science, but also to acknowledge the finite nature of that power and hold it in tension with our role in considering a person's wider well-being – i.e., health in its wider sense of wholeness and living a meaningful life.

Healthcare managers faced with the pressures of managing growing waiting lists with their limited resources might understandably favor technological solutions without understanding the deeper ramifications for staff and service users. Since the peak of the COVID-19 pandemic, the authors themselves have witnessed a pragmatic and swift normalization of remote telehealth musculoskeletal outpatient consultations, previously uncommon within the physiotherapy profession. The continuity and expansion of this approach offers clear benefits ranging from fulfillment of service demands with more appointment slots to supporting employees' preferences of working from home. However, it is too soon to tell what impact this change will have for patients and the wider physiotherapy profession, particularly in terms of the development of the next generation of physiotherapists as we take away the face-to-face, tactile human interaction. Working alone on a laptop is a sea change away from working in a bustling outpatient department. This might lead to feelings of isolation and loss of collegial spirit among fellow professionals so used to bouncing ideas off one another and challenging each other's practices. Without this in-person human interaction that many cite as the reason for doing the job, the authors' question whether this is still a role that they would be happy to fill? No matter what the data tell us, the inescapable fact is that most patients prefer to deal with another human being and anything that unnecessarily restricts that could fairly be considered a direction of travel away from person-centered practice. Furthermore, it stands to be seen what the longer-term data are on what has been missed in general practice in terms of important diagnoses since the shift to remote consultations.

AN EXAMPLE OF WHERE PEOPLE AND TECHNOLOGY MEET

Let us look at one example of the role of technology in supporting person-centered practice.

The COVID-19 pandemic led to a rapid transition to the use of digital technologies for remote healthcare, including physiotherapy. Restricted access to face-to-face therapy was important in reducing exposure and transmission of COVID-19. Existing community services were lacking in capacity to support those with long COVID. Telerehabilitation was piloted by a team of practitioners and researchers from a hospital in the North of England to support those who had been hospitalized with COVID-19 and had on-going symptoms, to help them on their path to recovery.

The telerehabilitation program was structured using conventional pulmonary rehabilitation principles with 12 sessions of group exercise, with educational sessions and peer support. It was delivered using a video conferencing application (Cisco WebEx Meetings, Cisco Systems Inc., USA). Prior to the start of the telerehabilitation program, a physiotherapist conducted a virtual initial assessment to check for eligibility, accessibility, and safety to exercise with remote supervision. Twice a week for six weeks, participants were invited to join a virtual group-based exercise program with three to six other people.

A qualitative evaluation was carried out to understand the views of patients hospitalized with COVID-19 on a telerehabilitation program. Participants reported that the telerehabilitation program helped them overcome some of the challenges they faced due to long COVID. For example, they perceived improvements in walking stamina, strength, and managing breathlessness, as well as improved sleep quality and fatigue management.

Despite never meeting physically face-to-face, the telerehabilitation program provided important social peer support. Those involved in the program found that, being with others who had been through a similar experience, was helpful and reduced some of the sense of isolation after returning home from hospital. These online social aspects helped them deal with some of the wider psychosocial challenges of their health and long-term recovery from COVID-19. Similarly, participants never met the physiotherapy instructors face-to-face. Yet, they reported that the physiotherapy instructors were central to the positive experience of the telerehabilitation program. Instructors were described as cheerful, positive, informative, full of joy, full of life, encouraging, and caring, with a cheery personality – they were perceived as being supportive and person-centered in their approach by treating participants as unique individuals.

We share this example to highlight the positive impact that can be realized when healthcare professionals bring person-centered practice and healthcare technology together. It shows that it is possible to cross that lay person/professional divide – we just need to be mindful of and positive toward that crossing. The telehealth program used technology to enhance human connection at a point in someone's life where they were at risk of social isolation. Good communication is at the heart of good healthcare, so we need to be considerate of how online support may help people connect with healthcare teams between visits. Thus, we need to think beyond efficiency, standardization, and protocol-driven outcomes and ensure that healthcare technology is applied in a people-centered way. We need to apply technology in a thoughtful way to ensure it supports patients as whole people with a lived body, and not just a physical body.

CONCLUSION AND FUTURE RECOMMENDATIONS

What we have sought to do with this chapter is to open a discourse on the interplay between emergent health technology and its human operators and recipients. From our perspective, one of the key questions is how we might go about harnessing health technology, with all its benefits, without losing the person-to-person contact we hold sacrosanct. It is vital to note that, just because we have a relational, person-to-person contact, this does not automatically constitute person-centeredness. Therefore, how can we ensure that this person-centered practice is taking place? Personally, we believe this is only possible if we have healthcare professionals who are willing and able to bridge the high-efficiency technological world that is modern healthcare to the world of the people with whom they interact. To be able to do this, practitioners must first acknowledge the high-technology environment in which they operate along with the invisible influences this can have on our clinical practice. Finally, they must show clinical bravery to be able to cross the boundary and enter the life world of the person, however uncomfortable this may be. Only then is it possible to deliver our truly person-centered vision of healthcare as it should be.

We offer the following reflective questions for healthcare professionals to consider:

- Which lens(es) do I use to view interactions with patients (scientific, biomedical, phenomenological, person-centered)?
- How do I view healthcare technology as part of my interactions with patients?
- What assumptions do I make about what is important to the person I am working with?
- Do I provide patients with an opportunity to share what is important or meaningful to them as part of healthcare interactions? In other words, how willing am I to cross into the world of the patient? What are the barriers that stop me doing this more?
- Do I then seek to incorporate what is important to them into their treatment plan?

REFERENCES

Al-Turjman, F., Hassan Nawaz, M. & Deniz Ulusar, U. 2020. Intelligence in the internet of medical things era: a systematic review of current and future trends. *Computer Communications*, 150, 644–660.

Berry, W. 1994. *Health is membership. Spirituality and healing.* Kentucky. http://tipiglen.co.uk/berryhealth.html

Borrell-Carrió, F., Suchman, A. & Epstein, R. 2004. The biopsychosocial model 25 years later: principles, practice, and scientific inquiry. *Annals of Family Medicine*, 2, 576–582.

Brun-Cottan, N., Mcmillian, D. & Hastings, J. 2020. Defending the art of physical therapy: expanding inquiry and crafting culture in support of therapeutic alliance. *Physiotherapy Theory & Practice*, 36, 669–678.

Brüssow, H. 2013. What is health? *Microbial Biotechnology*, 6, 341–348.

Busch, I., Moretti, F., Travaini, G., Wu, A. & Rimondini, M. 2019. Humanisation of care: key elements identified by patients, caregivers, and health care providers. A systematic review. *Patient*, 12, 461–474.

Charon, R. 2006. The self-telling body. *Narrative Inquiry*, 16, 191–200.

Engel, G. 1978. The biopsychosocial model and the education of health professionals. *Annals of the New York Academy of Sciences*, 310, 169–171.

Foot, C., Gilburt, H., Dunn, P., Jabbal, J., Seale, B., Goodrich, J., Buck, D. & Taylor, J. 2014. *People in control of their own health and care: the state of involvement*. London: The King's Fund.

Foster, N. & Delitto, A. 2011. Embedding psychosocial perspectives within clinical management of low back pain: integration of psychosocially informed management principles into physical therapist practice—challenges and opportunities. *Physical Therapy and Rehabilitation Journal*, 91, 790–803.

Gadamer, H. 1996. *The enigma of health: the art of healing in the scientific age*. Stanford: Stanford University Press.

Geisler, E. & Heller, O. 2012. *Managing technology in healthcare*. Boston: Kluwer Academic Publishers.

Goyen, M. & Debatin, J. 2009. Healthcare costs for new technologies. *European Journal of Nuclear Medicine and Molecular Imaging*, 36(Supplement 1), 139–143.

Groves, J. 2010. International alliance of patients' organisations perspectives on person-centred medicine. *International Journal of Integrated Care*, 10, 27–29.

Gwande, A. 2015. *Being mortal: illness, medicine and what matters in the end*. London: Profile books ltd.

Håkansson Eklund, J., Holmström, I., Kumlin, T., Kaminsky, E., Skoglund, K., Höglander, J., Sundler, A., Condén, E. & Summer Meranius, M. 2019. "Same same or different?" A review of reviews of person-centered and patient-centered care. *Patient Education and Counseling*, 102, 3–11.

Haraway, D. 2000. A manifesto for cyborgs: science, technology, and socialist feminism in the late 1980s. In Janes, L., Woodard, K. & Hovenden, F. (eds.), *The gendered cyboarg: a reader*. London: Open University.

Heidegger, M. 1977. *The question concerning technology and other essays*. New York: Harper and Row.

Holt-Lunstad, J., Smith, T. B. & Layton, J. B. 2010. Social relationships and mortality risk: a meta-analytic review. *Plos Medicine*, 7, e1000316–e1000316.

(IEEPO), I. E. E. F. P. O. 2021. *IEEPO 2021 position paper humanising healthcare: a call for transformational change*. https://www.ieepo.com/content/dam/websites/ieepo/2021/resources/IEEPO%20Position%20Paper_updated%20031121docx.pdf

Jacobs, G., Van Der Zijpp, T., Van Lieshout, F. & Van Dulmen, S. 2017. Research into person-centred healthcare technology. A plea for considering humanisation dimensions. In Mccormack, B., Van Dulmen, S., Eide, H., Skovdahl, K. & Eide, T. (eds.), *Person-centred healthcare research*. West Sussex: Wiley and Sons Ltd.

Jesus, T., Bright, F., Kayes, N. & Cott, C. 2016. Person-centred rehabilitation: what exactly does it mean? Protocol for a scoping review with thematic analysis towards framing the concept and practice of person-centred rehabilitation. *BMJ Open*, 6, 1–8.

Keller, K. 2020. The body as machine and the lived body in nursing. *Collegian*, 27, 238–244.

Lapum, J., Angus, J. E., Peters, E. & Watt-Watson, J. 2010. Patients' narrative accounts of open-heart surgery and recovery: authorial voice of technology. *Social Science & Medicine*, 70, 754–762.

Lapum, J., Fredericks, S., Beanlands, H., Mccay, E., Schwind, J. & Romaniuk, D. 2012. A cyborg ontology in health care: traversing into the liminal space between technology and person-centred practice. *Nursing Philosophy*, 13, 276–288.

Leder, D. 1984. Medicine and paradigms of embodiment. *The Journal of Medicine and Philosophy*, 9, 29–43.

Magee, B. 1987. *Descartes: a dialogue with Bernard Williams. The great philosophers: an introduction to western philosophy*. Oxford: Oxford University Press.

Mittelstadt, B. 2017. Ethics of the health-related internet of things: a narrative review. *Ethics and Information Technology*, 19, 157–175.

Mudge, S., Stretton, C. & Kayes, N. 2014. Are physiotherapists comfortable with person-centred practice? An autoethnographic insight. *Disability and Rehabilitation*, 36, 457–463.

NICE 2017. *Tailored resources: working with adults to ensure person-centred care and support for admissions*. London: National Institute for Health and Care Excellence.

Nicholls, D. A. & Gibson, B. E. 2010. The body and physiotherapy. *Physiotherapy Theory and Practice*, 26, 497–509.

O'Keefe-McCarthy, S. 2009. Technologically-mediated nursing care: the impact on moral agency. *Nursing Ethics*, 16, 786–796.

Roskell, C. 2013. An exploration of the professional identity embedded within UK cardiorespiratory physiotherapy curricula. *Physiotherapy*, 99, 132–8.

Sanders, T., Foster, N., Bishop, A. & Ong, B. 2013. Biopsychosocial care and the physiotherapy encounter: physiotherapists' accounts of back pain consultations. *BMC Musculoskeletal Disorders*, 14, 1–10.

Singh, R., Javaid, M., Haleem, A., Vaishya, R. & Ali, D. 2020. Internet of Medical Things (IoMT) for orthopaedic in COVID-19 pandemic: roles, challenges, and applications. *Journal of Clinical Orthopaedics and Trauma*, 11, 713–717.

Svenaeus, F. 2018. *Phenomenological bioethics. Medical technologies, human suffering, and the meaning of being alive*. London: Routledge.

Todres, L., Galvin, K. T. & Holloway, I. 2009. The humanization of healthcare: a value framework for qualitative research. *International Journal of Qualitative Studies on Health and Well-Being*, 4, 68–77.

Van Dulmen, S., Van Der Wees, P. & Nijhuis-Van Der Sanden, M. 2015. Patient-centered approach in clinical guidelines; a position paper of the guideline international network (G-I-N) allied health community. *Physiotherapy*, 101, E1575–E1576.

Vishnu, S., Jino Ramson, S. & Jegan, R. 2020. Internet of Medical Things (IoMT) - An overview. In *International conference on devices, circuits and systems*, Tamil Nadu, India.

WHO. 2015. *WHO global strategy on people-centred and integrated health services*. Geneva: World Health Organization.

WHO Interim Commission. (1948). *Official Records of the World Health Organization No. 2: Summary Report on Proceedings, Minutes and Final Acts of the International Health Conference Held in New York From 19 June to 22 July 1946*. Geneva: World Health Organization.

Wiles, R. & Barnard, S. 2001. Physiotherapists and evidence-based practice: an opportunity or threat to the profession? *Sociological Research Online*, 6, 62–74.

2 Augmentation through Technology
Considerations for Choice, Identity, and Culture

Charles Edmund Degeneffe

The modern day represents a period of extraordinary technological developments affecting all aspects of human life. For example, more than 90% of Americans have broadband access, a growth of 40% from the year 2000. More than half of the world's population enjoys Internet access. There are more cell phones (8 billion) than humans on Earth. Internet-based technologies are dramatically shaping many industries such as media, climate change efforts, and healthcare (Hillyer 2020). Likewise, the lives of persons with disabilities are being influenced by technological advancements and the adoption of new interventions, therapies, and functional supports.

Advances in technology have been a long-standing component of support for the needs of persons with disabilities in such domains as independent living, social life, and employment participation. Technology designed to enhance the lives of persons with disabilities is commonly referred to as "assistive technology" and is viewed as a way to equalize opportunities in the activities of daily life (Rubin and Roessler 1995) and maximize human potential (Power 2013). For many years (e.g., Rubin and Roessler 1995), rehabilitation counselors and other disability professionals have utilized a wide variety of technological supports to facilitate mobility, communication, health maintenance, cognitive and intellectual functioning, social and recreational activities, and activities of daily living among persons with disabilities.

Despite the availability of assistive technology, many people throughout the world lack access to such supports. Based on survey results from 29 countries, the World Health Organization and the United Nations International Children's Fund (UNICEF) (2022) estimated that more than 2.5 billion persons need some form of assistive technology. However, about one billion people lack access, especially in low- and middle-income countries. Survey results indicated that access was as low as 3% among those with reported technological needs (UNICEF 2022). To increase access to assistive technology, countries can develop targeted legislation and funding. For example, in the United States, the Assistive Technology Act of 2004 funds several programs and initiatives, including the State Grant for Assistive Technology Program (Administration for Community Living 2022). This program provides grants to both states and territories that "make assistive technology devices

DOI: 10.1201/9781003272786-2

and services more available and accessible to individuals with disabilities and their families."

As technology continues to advance, the discussion goes beyond expanding the capacity of technology to compensate for areas of disability that limit life participation to instead identifying how technology can actually reduce or even *reverse* disability. In discussing the potential impacts of biomedical engineering technology, Lee (2016, p. 69) refers to a future period of being "transhuman" or "posthuman," where humans could attain augmented abilities through technology. In athletics, Triviño (2013) used the term "cyborgs" (a combination of the words "cybernetic" and "organism") to characterize athletes whose function improved with the use of implanted robotic, electronic, or mechanical devices, in some cases resulting in superior performance compared with non-disabled competitors.

Because of the rapid pace of technological advances, the speed of its impacts on human activity often exceeds the ethical and moral considerations of its use. This caution applies to the consideration of technology to possibly reduce or reverse various types of cognitive, sensory, intellectual, and mobility impacts among persons with disabilities, which I refer to as "augmentative technology." I come to this topic from multiple perspectives. I am a Professor of Rehabilitation Counseling. Most of my scholarly work focuses on acquired brain injury in such areas as family caregiving, employment, and veteran issues. Before my academic career, I held a variety of positions working with individuals with disabilities in case management, supported employment, and residential direct care. These collective experiences provide an understanding of how persons with disabilities and their families find meaning, purpose, and identity with disabilities. My experiences also facilitate awareness of the challenges that disability can present to engaging in employment, relationships, and other domains of daily life.

This use of technology to ameliorate disability requires ethical considerations from the point of view of disabled persons in consideration of design, informed choice, and maintenance and respect of disability culture. This chapter will explore these areas and offer recommendations for continued discourse and debate.

USER-DIRECTED TECHNOLOGY

Assistive technology involves the use of "tools, equipment, or products" that can range from simple devices, like magnifying glasses or calculators, to complex machines and devices such as power wheelchair lifts and digital communication systems (Eunice Kennedy Shriver National Institute of Child Health and Human Development n.d.). Assistive technology today is transforming into more complex applications not previously thought possible. With contemporary advancements in genetics, robotics, nanotechnology, biomedicine, and robotics, the human body will become increasingly modifiable and will more directly interface with computers and physical structures designed to improve physical/sensory/cognitive function and performance (Triviño 2013). These emerging technologies could potentially augment human function beyond functional impairment or what would be considered typical human function. Two examples include cochlear implants (CI) and brain–computer interfaces (BCI).

One of the first approaches toward the use of implanted augmentative technology is CI. A CI device stimulates hearing through an implanted device that converts sounds into an electrical signal. Attempts to facilitate hearing through CI technology date back to 1748 with the introduction of extra-auricular electrical stimulation. Key steps in the development of modern-day CI technology included the first implanting of an electrode close to the vestibulocochlear nerve in 1957 and the insertion of the first commercially available multielectrode device in 1978 (Mudry and Mills 2013). According to the US National Institute on Deafness and Other Communication Disorders (2021), as of December 2019, approximately 736,900 CI devices had been implanted around the world. CI devices are designed to be "integrated into the functional world of the person" (Lee 2016, p. 70) in such areas as employment, education, and social participation.

Compared with the established use of CI, the development of BCI represents a more nascent form of augmentative technology. BCI comes from the discipline of neural engineering and refers to computational systems that place a person's central nervous system under the control of a device that improves autonomic, motor, and communication functions (Sullivan et al. 2018). BCI presents potential utility for several disability populations, including persons with spinal cord injury (SCI) (Collinger et al. 2013), amyotrophic lateral sclerosis (ALS) (Liberati et al. 2015), or a stroke (Yuan and He 2014). BCI devices are being designed in both non-invasive (e.g., placing sensors near or on top of the head to monitor brain activity) and invasive formats (i.e., implanting a device into the brain). Potentially, BCI technology could allow persons with disabilities to control devices simply through their thoughts and improve function beyond physical activity to also include enhanced cognitive skills with improved memory, attention, and overall cognitive performance (RAND Corporation 2020).

Beyond these forms of modern-day restorative and augmentative technologies, there are other examples. Triviño (2013) discussed the former track and field athlete Oscar Pistorius' use of carbon fiber prosthetics due to lower leg amputations of his legs, with some claiming Pistorius enjoyed a competitive advantage over other athletes. Smith (2022) described the ongoing development of bionic eyes to restore sight to persons with vision loss. Katoh and associates (2019) summarized past and current research focused on how to regenerate severed spinal cords through cell replacement and biomaterial implantation.

To non-disabled persons, these forms of technological advances could be viewed with excitement and the assumption that any therapeutic intervention that restores or improves function would naturally be accepted by those living with the impairments these technologies are designed to address. Attitudes like this reflect viewing disability through a medical model versus a social model. The medical model of disability equates disability with deficiency, where the assumption is that the person with a disability needs to be "fixed" or "cured" to achieve a meaningful quality of life and possess the capacity to participate in major life activities. The social model instead focuses on the need for society to modify its structures and institutions to facilitate participation in activities of daily life by disabled persons. An intervention from a social model perspective, for example, concerns the development of legislation that provides legal protections that allow individuals with disabilities to gain access to public services and places. For example, Australia's Disability Discrimination Act 1992 "makes it unlawful to discriminate against a person, in many areas of public life, including: employment, education, getting or using services, renting or buying a house or unit, and accessing public places, because of their disability" (Australian Human Rights Commission 2014, p. 1).

A central question in the development of technology designed to augment function concerns who is the intended beneficiary? It might be assumed this would naturally be the person with a disability. However, there could be other stakeholders that could possess different needs and motivations. For example, government agencies and/or insurance companies that fund disability services might want recipients of services and funding to adopt augmentative technologies to reduce their costs. Also, caregivers could want a family member with a disability to try a new intervention or therapy to increase their function, especially given the lack of services and funding that government systems allocate (e.g., Degeneffe 2019) to family caregiving worldwide. It is important to consider that persons with disabilities might present different needs and goals from other stakeholders and those who developed the technology, including choosing to live with their disability. In describing Hahn's (1991) critique of the functional limitation model of rehabilitation, Bedini (1993) stated, "…this model makes the incorrect assumption that people with disabilities wish to eliminate their disability. Hahn challenged this proposition by suggesting that some individuals with disabilities may find dignity and identity within their disability."

In recognition of the rights of persons with disabilities to define their needs and preferences for augmentative technology, several studies have obtained their perspectives. For instance, Collinger and colleagues (2013) surveyed 57 US veterans with SCI (37% with tetraplegia and 63% with paraplegia) on their views about BCI use. More than 80% of the sample endorsed BCI use if the technology did not present an inconvenience to other areas of their lives. Also, sample members conveyed the importance of the BCI device being non-invasive, allowing for independent use, and restoring muscle movement or function. Liberati and co-workers (2015) conducted a focus group on BCI use with nine persons in Italy with insider knowledge of ALS, including one individual with the condition along with relatives and professionals with ALS expertise. Focus group members stressed the need for more information on the capabilities and limitations of BCI technology, especially given media stories that "tend to present BCIs as devices that can 'read the mind,' fostering unrealistic expectations and rational skepticism" (p. 142). Focus group participants further suggested the BCI system should adapt to the needs of the person with ALS as the condition progresses.

Ultimately, the design of augmentative interventions relies on the perspective of the persons or systems that created the technology. The engineers, scientists, physicians, and therapists who design these technologies might lack a lived understanding of the intended recipients of the intervention and instead rely on their own professionally informed opinions. In the area of BCI, Kübler and associates (2014, p. 1) described this disparity as a "translational gap" between the perspectives of the developers of the technology and those of the intended users of their interventions. To demonstrate this point, Sullivan et al. (2018) interviewed 15 persons involved in the development of BCI technology representing the fields of electrical engineering, mechanical engineering, neurosurgery, computer science, rehabilitation medicine, physiology, neuroscience, and bioengineering. Participants indicated that there were benefits from engaging in direct experiences with persons with disabilities, while also gaining perspectives on the views of persons with disabilities through colleagues. Although there was encouragement for gaining the viewpoints of persons with disabilities in BCI development research, there was also doubt about how to accurately (e.g., with the wording of surveys) determine the views of potential users along with assessing the representativeness of their input when relying on small sample sizes or when users express contrasting opinions. For example, one participant stated the following:

> And so, it's, of course, vital to have end-user input, but to generalize from that enduser group to everyone, you have to be careful…I mean, what I take out of that is, okay, make sure that what we're doing is broad enough and flexible enough to satisfy what the enduser is saying they want.

(p. 489)

INFORMED CHOICE

A second consideration with the use of augmentative technologies concerns the importance of users making informed choices. The issue of informed consent and choice is a long-standing concern among persons with disabilities. There are countless examples (Braddock and Parish 2001) throughout human history of persons with disabilities being denied the opportunity to make informed decisions about the use of interventions, treatments, and services. For instance, on July 14, 1933, Germany passed the "Law for the Prevention of Progeny with Hereditary Disease," which required sterilization of all persons thought to possess heredity conditions including physical deformity, epilepsy, learning disabilities, mental illness, blindness, deafness, and advanced alcohol addiction (United States Holocaust Memorial Museum n.d.). Also, starting in 1955, Dr. Saul Krugman infected residents with intellectual disabilities (without their consent) at the Willowbrook State School in Staten Island, New York, with hepatitis for almost 20 years to better understand and develop treatments for the disease (Reimann 2017).

With the recognition of how disabled persons worldwide have lacked the right to make decisions regarding their medical care, services, and other life choices, the United Nations adopted the Convention on the Rights of Persons with Disabilities (and its Optional Protocol) on December 13, 2006, with 82 countries signing onto the Convention and 44 countries to the Optional Protocol. As of May 6, 2022, there are now 164 signers to the Convention and 94 to the Optional Protocol. To protect basic rights, the Convention articulates the following:

The Convention is intended as a human rights instrument with an explicit, social devel-opment dimension. It adopts a broad categorization of persons with disabilities and reaffirms that all persons with all types of disabilities must enjoy human rights and fundamental freedoms. It clarifies and qualifies how all categories of rights apply to persons with disabilities and identifies areas where adaptations have to be made for persons with disabilities to effectively exercise their rights and areas where their rights have been violated, and where protection of rights must be reinforced.

(United Nations 2022)

Despite this broad and powerful statement on protecting basic human rights, there is variability among nations, regional/province/state entities, and local articulations of how rights among persons with disabilities are defined and/or protected. For instance, in the United States, setting up guardianship for persons with disabilities varies among states (FindLaw 2023).

Hence, there exists the question of the degree of agency that persons with disabilities possess in deciding to use or not use an augmentative device. Persons with disabilities with court-appointed guardians or conservators may lack the legal authority to make this decision. The lack of agency also applies to children when parents are granted the legal authority to make medical and therapeutic decisions on behalf of their children. The lack of legal authority among many persons with disabilities to decide to use augmentative technology presents implications for fully understand-ing potential long-term benefits and risks. Fully understanding risks and benefits is especially pertinent to BCI technology, most of which remains at the early develop-ment stage. As noted, there are many potential physical and cognitive benefits of BCI devices. For example, future BCI devices might be able to help individuals with stroke, traumatic brain injury (TBI), or Alzheimer's disease improve memory function (RAND 2020), a benefit that could tremendously improve quality of life and the capacity to engage in different life activities. However, risks do exist. An implanted BCI device could result in brain damage, infection, or hemorrhaging, and the overall physical risks to the human body are not known. BCI devices could be used to facilitate external control of a person's thoughts, emotions, and actions. The long-term mental effects of BCI technology are also not yet known (RAND 2020). For instance, Solinsky and colleagues (2018) speculate about the potential for psychological harm in future BCI research for persons with SCI. Solinsky et al. (2018) noted that, in research with implanted BCI devices, persons with SCI would be informed that it could be removed at a future point due to possible infection or loss of technological functionality. If the BCI device resulted in gains in physical function and independence and was then removed, the psychological impacts of the immediate loss of these capabilities could be devastating to the research participants.

CI technology also presents both benefits and risks (US Food & Drug Administration 2021). CIs present the potential to facilitate overall hearing ability, the capacity to understand speech without reading lips, and the enhanced ability to engage in daily life activities like talking on the telephone, listening to music, and watching television. CIs can also assist users with perceiving different sound lev-els. By implanting a foreign device, however, there is the possibility of facial nerve damage, meningitis, cerebrospinal fluid leakage, perilymph fluid leak, skin wound infection, vertigo, tinnitus, taste disturbances, ear numbness, and reparative granuloma, among various other risks. Research by Dillon and Pryce (2020) suggests persons with hearing impairments weigh the risks versus the benefits of CI when deciding whether to use this technology. In their interviews with 15 deaf individuals in England possessing varying pre-study opinions and experiences with CIs, ranging from not wanting to use a CI to currently using a CI, Dillon and Pryce noted how risk considerations were evaluated:

Participants were aware of a wide range of risks including those resulting from surgery,

> health, unknown outcome with a CI, loss or residual hearing, and irreversibility, and linked these risks to their living context, future, and quality of life. A key component of decision making was a comparison exercise between the cost of impact of the risk compared with their current difficulties and the potential positive outcome they perceived that they may receive with CI, with mixed feelings over whether this was worth the risk.

(p. 28)

Related to informed choice is the principle of dignity. Meulen (2010) argues that all humans possess dignity regardless of their use or non-use of technology. Furthermore, in discussing the work of Jotterand (2010), Meulen warns of the danger of technology removing individuality by creating a type of standardization of human function that removes the individual differences that distinguish one person from another. The choices persons with disabilities will make about augmentative technology should consider how and if the ways users define themselves through disability would fundamentally change.

DISABILITY CULTURE

A third consideration of the use of augmentative technology is an acknowledgment of disability culture. Disability culture recognizes that persons with disabilities are part of a distinct cultural group that fights against oppression and denial of basic human rights. Disability culture promotes a positive and affirming sense of identity by seeing one's disability as a source of pride and self-worth. Finally, disability culture assumes a social justice stance against negative attitudes and socially created limits directed against disabled persons (Hopson 2019). Understanding disability culture is key to respecting how disabled persons view their place in the world (Ripat and Woodgate 2011).

A medical model view of disability might assume that *all* persons with disabilities would remove their disability if possible and regain areas of function lost or which were never attained. For non-disabled persons, it may be difficult, if not impossible, to understand how some persons view their disabilities as central to their identity and sense of purpose and cultural affiliation and would choose to not change their status. As Cherney (1999, p. 22) noted, "Finally, those who view disability as a lack, an insufficiency, or a challenge to be overcome often seek to 'solve' the 'problem' of disability through tech fixes." Disability can be key to developing a greater self-understanding of one's values and deeper insights into what is important in one's life. For example, in a qualitative study of 272 adult siblings of persons with TBI, Degeneffe and Olney (2010) reported that some participants found fundamental shifts in their values and priorities following their siblings' injuries, such as finding a deeper meaning in life, discovering new capacities, building closer relationships with family, and developing career goals informed by their family's disability experience. Also, through interviews with 11 persons with mild to profound hearing loss, Beckner and Helme (2018, p. 407) reported that most participants "...would not give up their hearing loss if they were granted the opportunity to have 'normal' hearing" since it was a core part of their identity.

The intersection of culture, disability, and technology is especially relevant for the deaf and hard-of-hearing community. Deaf culture distinguishes between the words "Deaf" and "deaf." The term "Deaf" refers to persons who are members of Deaf culture, made up of persons linked by a common language (e.g., American Sign Language), who hold common views and purposes about being deaf (Padden and Humphries 1988). The lowercase "deaf" is a broader term that indicates not being able to hear. An additional term is "hard of hearing" to include those with variations in hearing loss. The form of reference is a personal choice and reflects the diversity of the larger deaf community (National Association of the Deaf 2022). Members of Deaf culture advocate for home, work, and school environments that use vision as the main source of communication. Like other cultural groups, there is an established set of traditions, literature, artistic expression, and recreational and athletic activities that celebrate and strengthen Deaf culture (Laurent Clerc National Deaf Education Center 2015).

For some, the development of CI technology is a threat to the maintenance of Deaf culture, especially concerning CI use with children. In the United States, the US Food and Drug Administration (FDA) authorized the use of CIs for children over the age of two in 1990, a move that Cherney (1999, p. 28) referred to as "cultural genocide" in the Deaf community. One of the key objections to the FDA decision was removing from the children the choice to use a CI but giving this choice to their parents, most of whom were not deaf.

To gain a more nuanced and insider parental perspective, Hardonk and associates (2011) in Flanders, Belgium interviewed the parents of six deaf children (three to nine years old at the time of the interviews) with severe to profound hearing loss. Two children used a traditional hearing aid (THA) and a CI and the other four used a THA only. All but one of the fathers of the children were deaf. All of the mothers were deaf, with one using a CI. Except for the family with one hearing parent, the other participants "...gave priority to Deaf identity, sign language, and ethical issues in deciding between CI and hearing aids" (p. 290), with the authors concluding, "the decision-making processes of the parents involved factors that have also been found among hearing parents, as well as aspects that have not been reported to play in hearing parents' decision making" (p. 290). As an example of the concerns raised by the parents in the Hardonk et al. (2011) study, one father shared the following perspective on the potential loss of Deaf identity for children with a CI:

> Where are my Deaf friends? They have disappeared into the hearing world, where they have to make huge efforts to communicate with hearing people, which leaves them exhausted. Compared to when I was a child...we did everything with other Deaf people and it was fun!
>
> **(p. 297)**

Although some deaf individuals hold a high degree of identity and affiliation with Deaf culture, other deaf and hard-of-hearing individuals may lack this sense of belonging. For example, in Beckner and Helme's (2018) interviews with 11 persons with mild to profound hearing loss, participants expressed a sense of not fitting in with either the Deaf or the hearing community. One's sense of belonging and affiliation with Deaf culture can influence perceptions of using a CI device. For instance, Most and colleagues (2007) surveyed 115 deaf and hard-of-hearing adolescents with severe or profound hearing loss on their sense of Deaf identity and views toward CIs. The authors reported that participants with a strong Deaf identity had more negative views toward CI, while those with a bicultural identity (i.e., "I enjoy both deaf and hearing cultures," p. 73) reported more positive views. Given that deaf and hard-of-hearing persons hold varying degrees of Deaf identity, it is important to understand there is not just one view toward CI but rather multiple opinions. This point extends beyond the deaf and hard-of-hearing community to all persons with disabilities who will need to decide whether to use or not use augmentative technology in the future.

It is likely there will be debate and perhaps conflict among persons with disabilities and other stakeholders on the use of technology, regarding who speaks for disabled persons and the acceptable use of augmentative technology. To this point, Kirkham (2021) warns of the dangers of disability identity politics and critical disability studies, where he suggests the views of a minority of academics reflect their political views but may not represent the views of the larger community of persons with disabilities. In making this point, Kirkham shared:

> The identity politics practitioners are not content to just delay the provision of assistive technologies, by distracting from their production. They also wish to decide whether certain assistive technologies should be allowed to be used by other disabled people. In other words, they seek to take important decisions about disabled people's lives away from the individual and arrogate these decisions to themselves.
>
> **(p. 481)**

Ultimately, conversations about an individual disabled person's choice to use or not use technology should not occur in a vacuum, but rather with contextual considerations for age (Astell et al. 2020), culture, acculturation, gender, social class, and education (Ripat and Woodgate 2011).

CONCLUSION

Augmentative technology presents a new set of challenges and opportunities for persons with disabilities. These technologies hold promise for enhancing function and increasing participation in different domains of life. The development of these technologies is rapidly evolving. Their potential is hard to predict and only limited by the imaginations of the persons who conceptualize and develop new interventions, therapies, and supports.

A core theme of this chapter is honoring the adage, "Nothing about us without us." Too often, persons with disabilities have not been part of the conversation on the policies and programs developed to provide treatment and care. Throughout history, persons with disabilities have been subject to paternalistic attitudes about what is best for them, with too little examination of what persons with disabilities might want for themselves. Beyond lacking choice and freedom to define a future path, these attitudes have also resulted in systematic abuse, neglect, and exclusion.

These truths are especially important to remember in the continual development of augmentative technology. The purpose of this chapter was not to provide a prescriptive set of mandates or recommendations. Rather, I challenge those in positions of developing, implementing, and promoting new technologies to place persons with disabilities centrally in these processes. Persons with disabilities possess the right to define what they need, be informed about their options, and engage in self-examination on what technology means for their identity and place in the world.

REFERENCES

Administration for Community Living. (2022). *Assistive technology* [online]. Available at: https://www.unicef.org/reports/global-report-assistive-technology#:~:text=The %20WHO%2DUNICEF%20Global%20Report,that%20support%20communication %20and%20 cognition [accessed 11 July 2022].

Astell, A. J., McGrath, C. and Dove, E. (2020). 'That's for old so and so's!': Does identity influence older adults' technology adoption decisions? *Ageing & Society*, *40*(7), pp. 1550–1576.

Australian Human Rights Commission. (2014). *Disability discrimination* [online]. Available at: https://humanrights.gov.au/sites/default/files/GPGB_disability_discrimination.pdf [accessed 25 July 2022].

Beckner, B. N. and Helme, D. W. (2018). Deaf or hearing: A hard of hearing individual's navigation between two worlds. *American Annals of the Deaf*, *163*(3), pp. 394–412.

Bedini, L. A. 1993. Technology and people with disabilities: Ethical considerations. *Palaestra*, *9*(4), pp. 25–30.

Braddock, D. L. and Parish, S. L. (2001). An institutional history of disability. In Albrecht, G. L., Seelman, K. D. and Bury, M., eds. *Handbook of disability studies*. Thousand Oaks, CA: Sage Publications, pp. 11–68.

Cherney, J. L. (1999). Deaf culture and the cochlear implant debate: Cyborg politics and the identity of people with disabilities. *Argumentation and Advocacy*, *36*(1), pp. 22–34.

Collinger, J. L., Boninger, M. L., Bruns, T. M., Curley, K., Wang, W. and Weber, D. J. (2013). Functional priorities, assistive technology, and brain-computer interfaces after spinal cord injury. *Journal of Rehabilitation Research and Development*, *50*(2), p. 145.

Degeneffe, C. E. (2019). Understanding traumatic brain injury from a Gestalt approach. *Journal of Applied Rehabilitation Counseling, 50*(4), pp. 252–267.

Degeneffe, C. E. and Olney, M. F. (2010). 'We are the forgotten victims': Perspectives of adult siblings of persons with traumatic brain injury. *Brain Injury, 24*(12), pp. 1416–1427.

Dillon, B. and Pryce, H. (2020, January). What makes someone choose cochlear implantation? An exploration of factors that inform patient decision making. *International Journal of Audiology, 59*(1), pp. 24–32. doi: 10.1080/14992027.2019.1660917. Epub 2019 Sep 10. PMID: 31500481.

Eunice Kennedy Shriver National Institute of Child Health and Human Development. (n.d.). *Rehabilitative and assistive technology* [online]. Available at: https://www.nichd.nih.gov/health/topics/factsheets/rehabtech# [accessed 27 July 2022].

FindLaw. (2023). *State guardianship laws: Examples* [online]. Available at: https://www.findlaw.com/family/guardianship/examples-of-state-laws-on-guardianships.html [accessed 21 November 2023].

Hahn, H. (1991). Theories and values: Ethics and contrasting perspectives on disability. In Marinelli, R. P. and Dell Orto, A. E., eds. *The psychological and social impact of disability*. 3rd ed. New York: Springer Publishing Company, pp. 18–22.

Hardonk, S., Daniels, S., Desnerck, G., Loots, G., Van Hove, G., Van Kerschaver, E., Sigurjónsdóttir, H.B., Vanroelen, C. and Louckx, F. (2011). Deaf parents and pediatric cochlear implantation: An exploration of the decision-making process. *American Annals of the Deaf, 156*(3, Summer), pp. 290–304. doi: 10.1353/aad.2011.0027. PMID: 21941879.

Hillyer, M. (2020, 18 November). *How has technology changed-and changed us-in the past 20 years* [online]. Available at: https://www.themandarin.com.au/145425-how-has-technology-changed-and-changed-us-in-the-past-20-years/ [accessed 11 July 2022].

Hopson, J. (2019). Disability as culture. *Multicultural Education, 27*(1), pp. 22–24.

Jotterand, F. (2010). Human dignity and transhumanism: Do anthro-technological devices have moral status? *The American Journal of Bioethics, 10*(7), pp. 45–52.

Katoh, H., Yokota, K. and Fehlings, M. G. (2019). Regeneration of spinal cord connectivity through stem cell transplantation and biomaterial scaffolds. *Frontiers in Cellular Neuroscience, 13*, p. 248.

Kirkham, R. (2021). Why disability identity politics in assistive technologies research is unethical. In *Moving technology ethics at the forefront of society, organisations and governments*. Universidad de La Rioja, pp. 475–487. https://dialnet.unirioja.es/servlet/articulo?codigo=8037085

Kübler, A., Holz, E. M., Riccio, A., Zickler, C., Kaufmann, T., Kleih, S. C., Staiger-Sälzer, P., Desideri, L., Hoogerwerf, E. J. and Mattia, D., 2014. The user-centered design as novel perspective for evaluating the usability of BCI-controlled applications. *PloS one, 9*(12), p. e112392.

Laurent Clerc National Deaf Education Center. (2015). *American deaf culture* [online]. Available at: https://clerccenter.gallaudet.edu/national-resources/info/info-to-go/deaf-culture/american-deaf-culture.html [accessed 26 July 2022].

Lee, J. (2016). Cochlear implantation, enhancements, transhumanism and posthumanism: Some human questions. *Science and Engineering Ethics, 22*(1), pp. 67–92.

Liberati, G., Pizzimenti, A., Simione, L., Riccio, A., Schettini, F., Inghilleri, M., Mattia, D. and Cincotti, F. (2015). Developing brain-computer interfaces from a user-centered perspective: Assessing the needs of persons with amyotrophic lateral sclerosis, caregivers, and professionals. *Applied Ergonomics, 50*, pp. 139–146.

Meulen, R. T. (2010). Dignity, posthumanism, and the community of values. *The American Journal of Bioethics, 10*(7), pp. 69–70.

Most, T., Wiesel, A. and Blitzer, T. (2007). Identity and attitudes towards cochlear implant among deaf and hard of hearing adolescents. *Deafness & Education International*, *9*(2), pp. 68–82.

Mudry, A. and Mills, M. (2013). The early history of the cochlear implant: A retrospective. *JAMA Otolaryngology–Head & Neck Surgery*, *139*(5), pp. 446–453.

National Association of the Deaf. (2022). *Community and culture–frequently asked questions* [online]. Available at: https://www.nad.org/resources/american-sign-language/community-and-culture-frequently-asked-questions/ [accessed 26 July 2022].

National Institute on Deafness and Other Communication Disorders. (2021). *Cochlear implants* [online]. Available at: https://www.nidcd.nih.gov/health/cochlear-implants [accessed 26 July 2022].

Padden, C. A. and Humphries, T. (1988). *Deaf in America: Voices from a culture.* Cambridge, MA: Harvard University Press.

Power, P. (2013). *A guide to vocational assessment.* 5th ed. Austin, TX: PRO-ED.

RAND Corporation. (2020, 27 August). *Brain-computer interfaces are coming. Will we be ready?* [online]. Available at: https://www.rand.org/blog/articles/2020/08/brain-computer-interfaces-are-coming-will-we-be-ready.html [accessed 26 July 2022].

Reimann, M. (2017, 14 June). *Willowbrook, the institution that shocked a nation into changing its laws.* [online]. Available at: https://timeline.com/willowbrook-the-institution-that-shocked-a-nation-into-changing-its-laws-c847acb44e0d [accessed 27 July 2022].

Ripat, J. and Woodgate, R. (2011). The intersection of culture, disability and assistive technology. *Disability and Rehabilitation: Assistive Technology*, *6*(2), pp. 87–96.

Rubin, S. and Roessler, R. (1995). *Foundations of the vocational rehabilitation process.* 4th ed. Austin, TX: PRO-ED.

Smith, M. (2022, 10 March). *Bionic eyes: How tech is replacing lost vision* [online]. Available at: https://www.livescience.com/bionic-eye [accessed 27 July 2022].

Solinsky, R. and Specker Sullivan, L. (2018). Ethical issues surrounding a new generation of neuroprostheses for patients with spinal cord injuries. *PM&R*, *10*(Supplement 2), pp. S244–S248.

Sullivan, L. S., Klein, E., Brown, T., Sample, M., Pham, M., Tubig, P., Folland, R., Truitt, A. and Goering, S. (2018). Keeping disability in mind: A case study in implantable brain–computer interface research. *Science and Engineering Ethics*, *24*(2), pp. 479–504.

Triviño, J. L. P. (2013). Cyborgsportpersons: Between disability and enhancement. *Physical Culture and Sport*, *57*(1), pp. 12–21.

United Nations. (2022). *Convention on the rights of persons with disabilities (CRPD)* [online]. Available at: https://www.un.org/development/desa/disabilities/convention-on-the-rights-of-persons-with-disabilities.html [accessed 19 July 2022].

United States Children's Fund. (2022). *Global report on assistive technology: A WHO-UNICEF joint report* [online]. Available at: https://www.unicef.org/reports/global-report-assistive-technology#:~:text=The%20WHO%2DUNICEF%20Global%20Report,that%20support%20communication%20and%20 cognition. [accessed 11 July 2022].

United States Food and Drug Administration. (2021). *Benefits and risks of cochlear implants* [online]. Available at: https://www.fda.gov/medical-devices/cochlear-implants/benefits-and-risks-cochlear-implants [accessed 27 July 2022].

United States Holocaust Memorial Museum. (n.d.). *Nazi persecution of the disabled: Murder of the "unfit"* [online]. Available at: https://www.ushmm.org/information/exhibitions/online-exhibitions/special-focus/nazi-persecution-of-the-disabled [accessed 19 July 2022].

World Health Organization and the United Nations Children's Fund. (2022). *Global report on assistive technology.* License: CC BY-NC-SA 3.0 IGO.

Yuan, H. and He, B. (2014). Brain–computer interfaces using sensorimotor rhythms: Current state and future perspectives. *IEEE Transactions on Biomedical Engineering, 61*(5), pp. 1425–1435.

3 Emerging Technologies in Healthcare

Disability-Related Perspectives from Ghana

Augustina Naami and Vyda Mamly Hervie

INTRODUCTION – OVERVIEW OF PERSONS WITH DISABILITIES IN GHANA

Disability could be considered an unavoidable human circumstance because almost every individual is exposed to and is likely to experience disability on a temporary or permanent basis (Howard & Rhule, 2021). Persons with disabilities constitute about 15% of the world population, with about 80% of people with disabilities living in developing countries (World Health Organization, 2011). In Ghana, reports suggest that persons with disabilities represent about 8% (2,098,138) of the country's population of 30,832,019 (Ghana Statistical Service, 2021). It is worth noting that the number of persons with disabilities is likely to grow as incidences such as chronic ailments, falls, injuries, conflicts, and aging continue to occur (Howard & Rhule, 2021; World Health Organization [WHO], 2021).

In her efforts to demonstrate commitment to promoting and protecting the rights of persons with disabilities and facilitating inclusion and participation, Ghana passed the Persons with Disability Act (Act 715) in 2006. The Act protects the rights of persons with disabilities and makes specific provisions to guarantee their right to education, healthcare, employment, transportation, and community participation. Ghana has also ratified the Convention on the Rights of Persons with Disabilities (CRPD). Article 25 of this convention stipulates that "People with disabilities have the right to the enjoyment of the highest attainable standard of health without discrimination on the basis of disabilities" (United Nations, 2006, p. 18). Furthermore, Ghana adheres to the targets of the Sustainable Development Goals (SDGs), where universal access to healthcare is a key indicator of the performance of healthcare systems and a prerequisite to sustainable development. In this regard, access to healthcare services is seen as a measure of quality health outcomes (Ortega et al., 2018). Nonetheless, many persons with disabilities in Ghana remain disadvantaged in terms of adequate access to healthcare (Naami & Mfoafo-M'Carthy, 2020).

DOI: 10.1201/9781003272786-3

This chapter seeks to answer the following questions: How do persons with disabilities experience healthcare in Ghana? In what ways do the healthcare experiences of persons with disabilities in rural Ghana differ from their counterparts in urban centers? Guided by the social model of disability and the theory of intersectionality, we examine factors that could affect healthcare technology experiences of persons with various forms of disabilities, gender, age, and geographic locations.

THEORETICAL FRAMEWORKS

The theory of intersectionality and the social model of disability guided this study. Combining these theories as a framework aided in understanding how disability interacts with other factors to increase the healthcare vulnerabilities of persons with disabilities in Ghana. The theory of intersectionality was first developed in 1989 by a feminist legal scholar, Kimberle Crenshaw, who argued that single categorization of discrimination based on race, gender or any minority grouping could systematically omit the experiences of the more vulnerable groups. The proponents of intersectionality call for a re-examination of the definition of oppression to include the experiences of those who fall within multiple categories of vulnerabilities, such as disability, gender, age, and geographic region. Exploring the intersection of disability and other factors that contribute to healthcare vulnerabilities of persons with disabilities is critical to comprehensively understand the unique experiences of persons with disabilities relating to healthcare and emerging technologies. This knowledge will be useful for healthcare professionals, social workers, and professionals working in the area of disability and the government to develop appropriate policies and interventions that could address the healthcare needs of persons with disabilities arising from emerging technologies. The theory of intersectionality, however, is limited in deepening our understanding of how barriers such as social, physical, information, institutional, and transportation barriers affect access to benefits of emerging technologies. The social model of disabilities fills this gap.

The Social Model of Disability was proposed by advocates of the Union of the Physically Impaired Against Segregation (UPIAS) in 1976 (Shakespeare & Watson, 2002), which was later given credence by the works of scholars such as Mike Oliver (1990, 1997), Vic Finkelstein (1980, 1981), and Colin Barnes (1991) (Shakespeare & Watson, 2002). The proponents of the Social Model of Disability acknowledge that impairment could pose a limit to the functioning of persons with disabilities. However, they posit that persons with disabilities are disabled more by their environment and other processes than their perceived impairments (Union of the Physically Impaired Against Segregation [UPIAS], 1976). Examples of the disabling environment are societal structures, values, culture, environmental constructs (Geffen, 2013), inadequate access to healthcare, transportation, and physical and institutional barriers (Barnes & Mercer, 2005; Naami, 2014). The social model stresses the removal of barriers to foster the inclusion of persons with disabilities in mainstream society (Shakespeare & Watson, 2002; Naami, 2014). Applying the Social Model of Disability to this study helped us identify barriers that could inhibit access of persons with disabilities to emerging technologies and identify what could be done to break down those barriers.

HEALTHCARE IN GHANA

Access to healthcare for people in low- and middle-income countries can be challenging. Dynamics of access to healthcare in Africa include geographic location, accessibility, affordability, availability, and acceptability (Peters et al., 2008; O'Donnell, 2007).

Access to healthcare is a notable global challenge for many persons with disabilities, and this is worsened for persons with disabilities living in low- and middle-income countries, such as Ghana (Dassah et al., 2019). A report by the World Health Organization indicates that there is a high probability of persons with disabilities to be deprived of access to healthcare, as well as receive substandard care from medical professionals and hence encounter disparities in healthcare outcomes (WHO, 2021). Inadequate access to healthcare by persons with disabilities is attributed to factors such as poverty, inaccessible physical and transportation environments, and negative attitudes towards persons with disabilities and communication barriers (Ganle et al., 2016; Howard & Rhule, 2021; Naami, 2019; Vergunst et al., 2017; WHO, 2021).

Poverty is a major factor that hinders many persons with disabilities from accessing medical care in Ghana (Asuman et al., 2021; Abrokwah et al., 2020, Dassah et al., 2018; Drainoni et al., 2006; Mccoll et al., 2010; Rooy et al., 2012). Asuman et al. (2021), who sought to investigate disability and household welfare, concluded that 38.5% of households with persons with disabilities were below the national poverty line compared to 22.6% of households without persons with disabilities. Asuman et al. (2021) found a higher incidence of poverty in rural areas among households with persons with disabilities (51.5%) compared with households without persons with disabilities (38.7%).

Also, many hospital infrastructures, including entrances to the facilities, consulting rooms, laboratories, Out-Patient Departments (OPDs), and pharmacies, beds in admission and labor wards, laboratory and examination chairs and tables were not user-friendly for persons with visual or mobility disabilities (Abodey et al., 2020; Abrokwah et al., 2020; Badu et al., 2016). Further, available technology in healthcare settings has not been fully utilized to address the healthcare concerns of persons with disabilities because they are inaccessible. For example, hospital equipment such as scans, X-ray machines, and weighing and height measuring scales are not accessible to individuals with mobility disabilities (Abrokwah et al., 2020; Badu et al., 2016). Furthermore, attitudes (e.g., stigmatization, and abusive behaviors) of doctors, nurses, and laboratory technicians toward persons with disabilities were also reported to have a negative effect on access to healthcare and health outcomes for persons with disabilities (Abrokwah et al., 2020; Scheer et al., 2003; Amadhila, 2012; Howard & Rhule, 2021; Ganle et al., 2016).

Persons with disabilities in rural Ghana could encounter another layer of vulnerability because of the geographic region. Most rural areas in Ghana have poor road networks and inaccessible and expensive transportation systems, which makes it difficult to travel to urban areas to seek healthcare (Ganle et al., 2016; Naami & Mfoafo-M'Carthy, 2020; Novartis, 2022). Persons with visual and some forms of mobility impairment bear an additional cost of paying for personal assistant services (Abrokwah et al., 2020).

From the proceeding discussion, persons with disabilities in Ghana are likely to experience poor health outcomes (Frohmader & Ortoleva, 2014; Seidu et al., 2021). Those in rural Ghana could be even more vulnerable to poor quality of health. Telemedicine could be a panacea for addressing the many barriers that persons with disabilities face in accessing healthcare. It could improve both access to healthcare and quality of care.

EMERGING TECHNOLOGY

According to the Organization for Economic Co-operation and Development (OECD), health technology encompasses the application of practical and clinical knowledge to address health issues, including processes and practice mechanisms that change the modalities of healthcare delivery (OECD, 2017). Technological ecosystem in the health field plays a vital role in enhancing the living standards of persons with disabilities, assisting with diagnosis, and obtaining access to medication (WHO, 2022c). It is also useful in gathering adequate data for improving rehabilitative services for this vulnerable population (Elsaessar & Bauer, 2012). The relevance of technologies in healthcare for persons with disabilities is also due to the environmental and transportation barriers and ill-equipped facilities that impede access to healthcare for them (Abrokwah et al., 2020; Badu et al., 2016; Pugliese et al., 2022). Some advantages and disadvantages of telemedicine are listed in Table 3.1.

TABLE 3.1

Advantages, Disadvantages, and Possible Improvements Experienced by the Patient, Caregivers, and Clinical Staff during the Pilot Studies

Advantages	Disadvantages	Improvements
No travel	No personal touch	Better appointment options
In home	No emotional connection	Additional software options
Less time	Poor eye contact	Limit to necessary providers
Less effort	Video/audio	Better coordination of onsite support
Unaffected by bad weather	Wait time between providers	Better video/audio
More convenient	Lack of privacy	
Less stress	Not leaving house	
Seen by clinical staff	Coordination with onsite support	
No overnight stay	No physical examination	
No childcare worries		

Vasta et al. (2020)

The adoption of various technologies in the administration of healthcare is on the rise in sub-Saharan African countries, including Ghana (Holst et al., 2020). In Ghana, for instance, the government collaborated with the Novartis Foundation and other partners to pilot telemedicine in 30 communities in the Amansie West District of the Ashanti Region between 2011 and 2016 (Novartis, 2022). Also, in 2019, the government, in collaboration with the Zipline Company in the United States, started using drones to distribute blood and essential medicines in four locations in Ghana. When COVID-19 broke out, drones were used to carry blood samples across the operating areas (WHO, 2022b). Ghana also implemented the COVID-19 tracker application in 2020 to help track COVID cases, prevent the spread of the disease, and achieve better management of the pandemic (Government of Ghana, 2022). On July 19, 2022, Ghana launched its nationwide e-pharmacy platform to allow Ghanaians to buy their medicines using their mobile phones (Graphic Online, 2022). There are other telemedicine piloted projects in Ghana at various stages which are yet to be scaled up.

Notwithstanding the merits associated with healthcare technologies (see Table 3.1), the full utilization of technology to deliver health services for the benefit of persons with disabilities is yet to be realized. The subsequent section highlights some issues related to the use of healthcare technology by persons with disabilities in Ghana.

ISSUES RELATING TO THE USE OF TELEMEDICINE FOR PERSONS WITH DISABILITIES IN GHANA

Using digital platforms could facilitate ease of access to health services for persons with disabilities in Ghana, given that many barriers impede their access to healthcare. However, in Ghana, many telemedicine projects are still at the pilot stage. The few that exist require the use of Android phones and internet services. However, internet data costs are relatively high in Ghana (Mensah, 2021) and many persons with disabilities cannot afford Android mobile phones owing to poverty (Naami & Mfoafo-M'Carthy, 2020), the major factor that hinders many persons with disabilities from accessing formal medical care (Abrokwah et al., 2020, Dassah et al., 2018; Drainoni et al., 2006; Mccoll et al., 2010; Rooy et al., 2012). Poverty could deter many persons with disabilities in Ghana from using relevant technologies, such as e-pharmacy, to address their health needs.

Persons with disabilities in rural Ghana, where poverty is prevalent (Ghana Statistical Service [GSS], 2018), could even be more vulnerable due to geographic location. In addition to the difficulty of paying for Android phones and internet services, some of these areas are yet to be connected to the national electricity grid. All these factors could complicate the experiences of persons with disabilities in rural Ghana. Women with disabilities are also likely to be affected adversely because of the interaction of gender and disability. Women with disabilities are not only more likely to be poorer than their counterparts without disabilities, but also than their male counterparts (Mitra, 2006; Naami, 2015; WHO, 2011). Furthermore, disability and old age could influence the experiences of older people with disabilities, as well as older women with disabilities.

In addition, to obtain good outcomes, telemedicine must be user-friendly to all. However, communication barriers could affect the use of emerging technology for persons with hearing and/or speech impairment. Currently, there is a lack of Sign Language interpreters at healthcare institutions, making it difficult for persons with hearing impairment to communicate with healthcare professionals (Abrokwah et al., 2020; Baart & Taaka, 2017; Mprah, 2013). Some literate individuals with hearing impairment use note-writing to communicate with healthcare professionals (Abrokwah et al., 2020). Will the government of Ghana and the management of healthcare institutions be willing to add devices such as Telecommunication Device (TDD) and text telephone (TTY) for the deaf, which usually comes across at an additional cost? There is another group of persons with hearing impairment who are not literate and cannot communicate in Sign Language (Abrokwah et al., 2020). How could telemedicine be designed to reach this group of people? In addition, are those with visual impairment, who may have access to Android phones but can either not afford to pay for screen readers or are not literate, able to use technology.

Furthermore, to use emerging technology requires knowledge of the use of the technology by both the professionals and the beneficiaries (Pugliese et al., 2022). The beneficiaries must read and understand what is required to effectively use each technology. But lower educational attainment could affect the usage of telemedicine by persons with disabilities. The World Disability Report indicated that many people with disabilities across the globe have lower educational attainment (WHO, 2011). How many persons with disabilities in Ghana can read and understand instructions for using technology? Will healthcare professionals be willing to help persons with disabilities to use telemedicine? Will they be patient enough to explain the use of technology to persons with disabilities or adjust them to address their unique needs, given that negative attitudes of healthcare professionals towards persons with disabilities are still rife? Working from home requires equipment and other necessary support available to workers with disabilities at their workplaces, such as screen readers for the blind.

CONCLUSION

Telemedicine is and will remain an essential component of the global healthcare system, given its relevance in enhancing healthcare (Bhatia, 2021; Vasta et al., 2020). Emerging healthcare technology could undoubtedly improve healthcare for persons with disabilities in Ghana, whose experiences are characterized by several barriers, including social, physical, information, communication, and transportation. Although various technologies have been introduced to improve healthcare service delivery in other countries, it is not the same for Ghana. Factors such as economic, cultural, physical, communication, and information accessibility issues could affect the access of persons with disabilities to the few telemedicine programs in Ghana. Disability type, gender, age, and geographic region could further affect the experiences of persons with disabilities who fall within these socio-demographic dimensions.

As Ghana continues to explore telemedicine programs, it is prudent to consider the unique needs of persons with disabilities, especially those who fall within multiple vulnerabilities, such as women and elderly persons with disabilities from rural areas. The government should ensure that emerging technology is accessible to all, including persons with disabilities. Since access to technology also depends on one's ability to pay, the government could consider subsidizing the cost for persons with disabilities who are likely to have many health needs. By doing this, healthcare inequality for persons with disabilities could be reduced. It is ethically wrong to withhold services from persons with disabilities because they cannot afford them. It is equally important for Ghana to adhere to Article 25 of the Convention on the Rights of Persons with Disabilities, which requires state parties to develop measures to ensure that persons with all forms of disabilities enjoy the highest attainable healthcare without discrimination.

Furthermore, Ghana could adhere to the targets of the Sustainable Development Goals (SDGs), where universal access to healthcare is a key indicator of the performance of healthcare systems and a prerequisite of sustainable development. To improve healthcare for persons with disabilities in Ghana, the government, the ministry in charge of healthcare, the healthcare directorate, and healthcare professionals should collaborate with organizations representing persons with disabilities to address the attitudes of healthcare professionals toward persons with disabilities.

REFERENCES

Abodey, E., Vanderpuye, I., Mensah, I. and Badu, E., 2020. In search of universal health coverage–highlighting the accessibility of health care to students with disabilities in Ghana: A qualitative study. *BMC Health Services Research*, *20*(1), pp. 1–12.

Abrokwah, R., Aggire-Tettey, E. M. and Naami, A., 2020. Accessing healthcare in Ghana: Challenges encountered and strategies adopted by persons with disabilities in Accra. *Disability, CBR & Inclusive Development*, *31*(1), 120–141.

Amadhila, E., 2012. *Barriers to accessing health care for the physically impaired population in Namibia* (Doctoral dissertation).

Asuman, D., Ackah, C. G. and Agyire-Tettey, F., 2021. Disability and household welfare in Ghana: Costs and correlates. *Journal of Family and Economic Issues*, *42*(4), pp. 633–649.

Baart, J. and Taaka, F., 2017. Barriers to healthcare services for people with disabilities in developing countries: A literature review. *Disability, CBR & Inclusive Development*, *28*(4), pp. 26–40.

Badu, E., Agyei-Baffour, P. and Opoku, M. P., 2016. Access barriers to health care among people with disabilities in the Kumasi Metropolis of Ghana. *Canadian Journal of Disability Studies*, *5*(2), pp. 131–151.

Barnes, C., 1991. *Disabled people in Britain and discrimination: A case for anti-discrimination legislation*. London: Hurst Co.

Barnes, C. and Mercer, G., 2005. Disability, work, and welfare: Challenging the social exclusion of disabled people. *Work, Employment and Society*, *19*(3), pp. 527–545.

Bhatia, R., 2021. Emerging health technologies and how they can transform healthcare delivery. *Journal of Health Management*, *23*(1), pp. 63–73.

Dassah, E., Aldersey, H. M., McColl, M. A. and Davison, C., 2018. Factors affecting access to primary health care services for persons with disabilities in rural areas: a "best-fit" framework synthesis. *Global health research and policy*, *3*(1), pp. 1–13. https://doi.org /10.1017/S1463423619000495.

Drainoni, M. L., Lee-Hood, E., Tobias, C., Bachman, S. S., Andrew, J. and Maisels, L., 2006. Cross-disability experiences of barriers to health-care access: Consumer perspectives. *Journal of Disability Policy Studies*, *17*(2), pp. 101–115.

Elsaesser, L. J. and Bauer, S., 2012. Integrating medical, assistive, and universal design prod-uctsand technologies: Assistive Technology Service Method (ATSM). *Disability and Rehabilitation: Assistive Technology*, *7*(4), pp. 282–286.

Finkelstein, J.J., 1981. The ox that gored. *Transactions of the American Philosophical Society*, *71*(2), pp. 1–89.

Finkelstein, V., 1980. *Attitudes and disabled people: Issues for discussion (No. 5)*. World Rehabilitation Fund, Incorporated.

Ganle, J. K., Otupiri, E., Obeng, B., Edusie, A. K., Ankomah, A. and Adanu, R., 2016. Challenges women with disability face in accessing and using maternal healthcare ser-vices in Ghana: A qualitative study. *PloS one*, *11*(6), p. e0158361.

Geffen, R., 2013. The equality Act 2010 and the social model of disability. p. 2017. http://pf7 d7vi404s1dxh27mla5569.wpengine.netdnacdn.com/files/library/THE%20EQ202010 %20%20THE%20SOCIAL%20MODEL%20TO%20USE.pdf [accessed 4 June 2017].

Ghana Statistical Service, 2018. *Ghana Living Standards Survey Round 7 (GLSS 7): Poverty trends in Ghana 2005–2017*. Accra, Ghana.

Ghana Statistical Service, 2021. *Ghana 2021 population and housing census*. https://census2021.statsghana.gov.gh/

Government of Ghana-Ministry of Health, 2022. *Launch of GH COVID-19 tracker app ministry of communications* [Internet]. https://www.moc.gov.gh/launch-gh-covid-19 -tracker-app [cited 23 August 2021].

Graphic Online. 2022. *Ghana launches national-scale ePharmacy platform*. https://www .graphic.com.gh/news/health/ghana-launches-national-scale-epharmacyplatform.html

Holst, C., Sukums, F., Radovanovic, D., Ngowi, B., Noll, J. and Winkler, A. S., 2020. SubSaharan Africa—The new breeding ground for global digital health. *The Lancet Digital Health*, *2*(4), pp. e160–e162.

Howard, H. A. and Rhule, A. B., 2021. Socioeconomic factors hindering access to healthcare by persons with disabilities in the Ahanta West Municipality, Ghana. *Disability, CBR &Inclusive Development*, *32*(2), p. 69.

McColl, M. A., Jarzynowska, A. and Shortt, S. E. D., 2010. Unmet health care needs of people with disabilities: Population level evidence. *Disability & Society*, *25*(2), pp. 205–218.

Mensah, C., 2021. Using web-survey to collect data on psychological impacts of COVID-19 on hotel employees in Ghana: A methodological review. *Cogent Psychology*, *8*(1), p. 1880257.

Mitra, S., 2006. The capability approach and disability. *Journal of Disability Policy Studies*, *16*(4), pp. 236–247.

Mprah, W. K., 2013. Perceptions about barriers to sexual and reproductive health information and services among deaf people in Ghana. *Disability, CBR & Inclusive Development*, *24*(3), pp. 21–36.

Naami, A., 2014. Breaking the barriers: Ghanaians' perspectives about the social model. *Disability, CBR & Inclusive Development*, *25*(1), pp. 21–39.

Naami, A., 2015. Disability, gender, and employment relationships in Africa: The case of Ghana. *African Journal of Disability*, *4*(1), pp. 1–11.

Naami, A., 2019. Access barriers encountered by persons with mobility disabilities in Accra, Ghana. *Journal of Social Inclusion*, *10*(2), pp. 26–46.

Naami, A. and Mfoafo-M'Carthy, M., 2020. COVID-19: Vulnerabilities of persons with disabilities in Ghana. *African Journal of Social Work*, *10*(3), pp. 9–17.

Novartis, 2022. Ghana Telemedicine. Available Online from https://www.novartisfoundation .org/past-programs/digital-health/ghana-telemedicine

O'Donnell, O., 2007. Access to health care in developing countries: Breaking down demand side barriers. *Cadernos de saude publica*, *23*, pp. 2820–2834.

OECD, 2017. *New health technologies: Managing access, value and sustainability*. OECD Publishing. https://doi.org/10.1787/9789264266438-en

Oliver, M., 1990. Social policy and disability: Some theoretical issues. *Disability Handicap, and Society*, *1*(1), pp. 5–18.

Oliver, M., 1997. The disability movement is a new social movement! *Community Development Journal*, *32*(3), pp. 244–251.

Ortega, A. N., McKenna, R. M., Pintor, J. K., Langellier, B. A., Roby, D. H., Pourat, N., Bustamante, A. V. and Wallace, S. P., 2018. Health care access and physical and behavioral health among undocumented Latinos in California. *Medical care*, *56*(11), p. 919.

Peters, D. H., Garg, A., Bloom, G., Walker, D. G., Brieger, W. R. and Hafizur Rahman, M., 2008. Poverty and access to health care in developing countries. *Annals of the New York Academy of Sciences*, *1136*(1), pp. 161–171.

Pugliese, R., Sala, R., Regondi, S., Beltrami, B. and Lunetta, C., 2022. Emerging technologies for management of patients with amyotrophic lateral sclerosis: From telehealth to assistive robotics and neural interfaces. *Journal of Neurology*, *269*(6), pp. 2910–2921.

Seidu, A. A., Malau-Aduli, B. S., McBain-Rigg, K., Malau-Aduli, A. E. and Emeto, T. I., 2021. Level of inclusiveness of people with disabilities in Ghanaian health policies and reports: A scoping review. *Disabilities*, *1*(3), pp. 257–277.

Scheer, J., Kroll, T., Neri, M. T. and Beatty, P., 2003. Access barriers for disabled people: The consumer's perspective. *Journal of Disability Policy Studies*, *13*, pp. 221–230.

Shakespeare, T. and Watson, N., 2002. The social model of disability: An outdated ideology? In S. Barnarrt and B. M. Altman, eds. *Exploring theories and expanding methodologies: Where are we and where do we need to go? Research in social sciences and disability*, Vol. 2. Amsterdam, JAI.

Union of the Physically Impaired Against Segregation. 1976. Fundamental principles of disability. Available Online from https://disability-studies.leeds.ac.uk/wp-content/uploads /sites/40/library/UPIAS-fundamental-principles.pdf [Accessed on 17 June 2017]

United Nations. 2006. *Convention on the rights of persons with disabilities.* Available Online from http://www.un.org/disabilities/documents/convention/convoptprot-e.pdf

Van Rooy, G., Amadhila, E. M., Mufune, P., Swartz, L., Mannan, H. and MacLachlan, M., 2012. Perceived barriers to accessing health services among people with disabilities in rural northern Namibia. *Disability & Society*, *27*(6), pp. 761–775.

Vasta, S., Papalia, R., Torre, G., Vorini, F., Papalia, G., Zampogna, B., Fossati, C., Bravi, M., Campi, S. and Denaro, V., 2020. The influence of preoperative physical activity on postoperative outcomes of knee and hip arthroplasty surgery in the elderly: A systematic review. *Journal of Clinical Medicine*, *9*(4), p. 969.

Vergunst, R., Swartz, L., Hem, K. G., Eide, A. H., Mannan, H., MacLachlan, M., Mji, G., Braathen, S. H. and Schneider, M., 2017. Access to health care for persons with disabilities in rural South Africa. *BMC Health Services Research*, *17*(1), pp. 1–8.

World Health Organization, 2011. *World report on disability 2011.* World Health Organization.

World Health Organization, 2021. *Disability and health.* World Health Organization. Available Online from https://www.who.int/news-room/fact-sheets/detail/disability -and-health

World Health Organisation, 2022a. Disability and health. Available Online from https://www .who.int/news-room/fact-sheets/detail/disability-and-health

World Health Organisation, 2022b. *His excellency the vice president of the republic of Ghana launches Ghana's drone delivery service.* Available Online from https://www.afro.who.int/news/his-excellency-vice-president-republic-ghana-launches-ghanas-drone-delivery-service

World Health Organization, 2022c. *Global report on health equity for persons with disabilities.* World Health Organization.

4 Emerging Technologies in Neurorehabilitation
A Perspective from Brazil

Christina Danielli Coelho de Morais Faria, Aline Alvim Scianni, Paula da Cruz Peniche, Sherindan Ayessa Ferreira de Brito, and Luci Fuscaldi Teixeira-Salmela

INTRODUCTION

Neurological conditions represent a heavy burden worldwide (PAHO, 2021). Over recent years, the prevalence and incidence of neurological conditions in most parts of the world have stabilized or decreased. However, in Brazil, both the prevalence and incidence of such conditions have grown significantly (PAHO, 2021). Neurological conditions account for 1,242.50 disability-adjusted life years (DALY) and an average of 813.7 years lived with disability (YLD) related to these conditions per 100,000 of the population. These levels of burden are ranked highest among all of the countries in the world (PAHO, 2021). This has resulted in increases in the health system demands, which justifies the need to strengthen neurorehabilitation services in Brazil.

Emerging technologies may potentially improve delivery of neurorehabilitation services. This is especially important in Brazil, which is a developing country with a large population and high demand for health services. Nowadays, Brazil is classified as a middle-income country that has high sincome inequality. Approximately 70% of the population depends exclusively on the services and actions of the Brazilian public health system, named the Unified Health System (Araújo et al., 2017). In this scenario, the use of emerging technologies in the context of neurorehabilitation is an important strategy to improve the provision of healthcare services. On the other hand, due to geographic, economic, social, cultural, and educational challenges, historically found in Brazil, the use of these technologies is still uncommon.

It has been reported that racial discrimination is a structured contributor to economic and social disadvantages experienced by racial/ethnic minorities (Hasenbalg & Silva, 1999). In addition, race has been suggested as a predictor of mortality in Brazil. Early mortality is more frequent among Indigenous and Black people, and mortality rates related to a stroke are much higher among Black women. Furthermore, lifetime socioeconomic differences across successive generations have

DOI: 10.1201/9781003272786-4

been identified as the main cause of racial inequality in healthcare (Chor & Lima, 2005). However, racial inequalities in health have received little attention. Therefore, the use of emerging technologies in Brazil should be discussed and considered in terms of foreseeable economic, social, cultural, and ethnicity-related issues.

Worldwide, emerging technologies in the context of neurorehabilitation have been used for a variety of purposes (Brennan et al., 2009), such as to assist in diagnosis, prognosis, and clinical decision-making, as well as monitoring and evaluating intervention effects. In addition, telehealth and/or telerehabilitation have been used to assess, monitor, and treat individuals remotely (Brennan et al., 2009, Brito et al., 2022; Peretti et al., 2017). Therefore, these technologies have the potential to facilitate and improve rehabilitation of individuals with neurological conditions. However, for their use within clinical contexts, several aspects should be considered, such as cost, accessibility, feasibility, and safety (Brito et al., 2020).

In many countries, emerging technologies have been increasingly employed in the context of neurorehabilitation, both within research and clinical contexts. Important clinical guidelines have recommended the use of technologies, such as virtual reality therapy and robotics. For instance, the Guidelines for Adult Stroke Rehabilitation and Recovery from the American Heart Association/American Stroke Association recommended the use of virtual reality for delivering upper-extremity movement practice and improving visual-spatial/perceptual functioning. The use of robotic devices for mechanically assisted walk training has been also recommended (Winstein et al., 2016). A clinical practice guideline for physical therapy management of Parkinson's disease from the American Physical Therapy Association suggested the use of virtual reality, robotics, and other technologies to support the delivery of rehabilitation (Osborne et al., 2022).

Unfortunately, recommendations on the use of emerging technologies were not found in the main Brazilian clinical guidelines directed at neurorehabilitation, since the guidelines or protocols aimed at rehabilitating individuals suffering from a stroke, Parkinson's disease, and amyotrophic lateral sclerosis do not address this issue (Brasil, 2013, 2017, 2020), which makes it difficult for clinicians to access information and use these technologies. Brazilian people are Portuguese speakers and one important barrier faced by neurorehabilitation professionals is difficulties in reading and understanding the English language (Nascimento et al., 2020). Therefore, Brazilian clinical guidelines are more applicable and more often accessed by those professionals than international ones. Whereas in many countries these technologies have been increasingly used in the context of neurorehabilitation, in Brazil, they have not even been mentioned in the main references developed to drive clinical decision-making. Why does it happen?

CHALLENGES RELATED TO THE USE OF EMERGING TECHNOLOGIES IN THE CONTEXT OF NEUROREHABILITATION IN BRAZIL

One important challenge is related to the country's large geographic dimensions. Brazil is the largest country in South America and the fifth largest on the planet, with a population of approximately 213 million people (Figure 4.1). It is composed of five regions (Midwest, Northeast, North, Southeast, and South) with 27 federation units. These regions represent important economic, social, cultural, and educational inequalities. For instance, in the Southeast region, the most populous region, with the highest gross domestic product, 34.9% of the population has access to private health insurance, whereas, in the North region, only 13% has (IBGE, 2020). Due to these inequalities, the country faces important challenges in implementing emerging technologies within clinical contexts (Table 4.1).

Another important challenge is related to infrastructure and costs (Anwar & Shamim, 2011). It is necessary to have an organized infrastructure to incorporate these technologies into clinical practice and achieve expected goals. This requires financial, structural, and human resources, such as the provision of computer hardware and software, internet availability, and qualified professionals, but this is difficult for low- and middle-income countries, such as Brazil (Anwar & Shamim, 2011). These countries commonly have poor or inadequate infrastructure, inadequate provision of computer hardware and software, and lack of qualified human resources personnel and training to qualify professionals, as well as poor internet availability (Anwar & Shamim, 2011). This is even more challenging, due to the country's large geographical area and population and its inequalities. Brazil is one of the most unequal countries in the world, with a Gini index of 0.543. The Gini index is an important indicator of inequality that allows for international comparisons (IBGE, 2020). The average monthly income is about US$470 (IBGE, 2020) and only 55.6% of the households located in the rural areas have access to the internet (IBGE, 2019). These numbers are even lower in some regions, such as the rural areas of the North region, where only 38.4% of households have internet access (IBGE, 2019). Therefore, access to technologies is restricted for a high proportion of

FIGURE 4.1   Location of Brazil.

TABLE 4.1

Main Challenges Related to the Use of Emerging Technologies in Brazil.

- Infrastructure
  - Lack of financial, structural, and qualified human resources
- Cost
- Lack of access for both patients and professionals
- Educational, social, and ethnic issues
- Difficulty in using available scientific evidence
  - Difficulty in obtaining full-text papers
  - Cost
  - Language of publication of the papers
  - Lack of quality evidence
  - Lack of training in evidence-based practice
  - Insufficient time provided by management
  - Lack of generalizability of research findings to the patient population
  - No deployment of scientific research
  - Lack of understanding of statistical analysis
- Organizational challenges
  - Lack of literature on costs, risk analysis, and other aspects of managing emerging health technologies in underdeveloped and developing countries

the Brazilian population. Health professionals also face financial difficulties to seek training, workshops, and other sources of knowledge in this area. In addition, there is a poor incentive and investment to train these professionals in the use of those technologies, since the Brazilian public healthcare system faces problems that are considered priorities, such as immunization programs, expansion of basic sanitation and environmental health, and reduction of risks and harm to the health of the population, through actions of health promotion, prevention, and surveillance (Conselho Nacional de Saúde, 2022).

Educational and social issues are also important challenges. The Brazilian population has a significant illiteracy rate and inequality in education levels. For instance, 18% of people over 60 years of age are illiterate. In addition, 44.3% have only completed elementary education, 41.2% have incomplete secondary to incomplete higher education, and only 14.5% have completed higher education (IBGE, 2020). Therefore, many patients do not have the knowledge and face difficulties in using these technologies.

Another important challenge refers to the difficulty that Brazilian health professionals have in applying evidence-based practice in their clinical practice. Evidence-based practice combines the best available scientific evidence, professional expertise, and patient preference in clinical decision-making, which is important for the proper use of emerging technologies within clinical contexts. Previous studies identified several barriers to the use of evidence-based practice in Brazil that prevented the use of these technologies, such as difficulty in obtaining full-text papers, cost, language of the papers, lack of quality evidence, lack of training (Silva et al., 2015), insufficient time provided by management, lack of generalizability of research findings to the patient population, no deployment of scientific research, and lack of understanding of statistical analysis (Nascimento et al., 2020).

Finally, all these issues lead to organizational challenges, that make it difficult to implement these technologies in clinical practice. Literature on costs, risk analysis, and other aspects of management is scarce, especially in low- and middle-income countries. Therefore, managers of health organizations in Brazil have difficulties in implementing emerging technologies, especially in public institutions (Anwar & Shamim, 2011).

TIME TO ACT AND ADVANCE ON THE USE OF EMERGING TECHNOLOGIES IN BRAZIL

Some efforts have been made in attempting to advance the provision of access to health technologies, in general, which may reflect in improved access to emerging technologies for the neurorehabilitation scenario. The *Comissão Nacional de Incorporação de Tecnologias no Sistema Único de Saúde* (CONITEC) (in English, the National Commission for the Incorporation of Technologies in the Unified Health System) was created in 2011 (Comissão nacional de incorporação de tecnologias no Sistema Único de Saúde, 2021). The objective of this committee is to advise the Minister of Health on the attributions related to the incorporation, exclusion, or changes in health technologies by the Unified Health System, as well as the elaboration or changes in clinical protocols or therapeutic guidelines. However, the incorporation of emerging technologies into the health technologies already recommended by CONITEC is still scarce. In addition, the clinical protocols or therapeutic guidelines for the care of individuals with neurological conditions provided by CONITEC do not include these technologies and are mostly related to the use of drugs and surgical procedures. This confirms the need for actions aimed at including emerging technologies, which would allow professionals working in public health services to have access to these technologies in the rehabilitation scenario of individuals with neurological conditions. In addition, the clinical protocols or therapeutic guidelines provided by CONITEC need to be updated to incorporate the use of emerging technologies in the context of neurorehabilitation, to be in line with international clinical guidelines.

Efforts have also been made to increase access to information for both clinicians and patients. Research groups from several universities have carried out the dissemination of studies and publications in various social media. In addition, several Brazilian institutions and associations have also made efforts to disseminate information, by holding online events, in which the use of emerging technologies in neurorehabilitation has been addressed. These events are freely accessible to patients, students, and clinicians. These actions have allowed for free and easy access to information in an accessible language. All of these actions have the potential to increase access to and disseminate evidence on the use of emerging technologies.

Nowadays, access to emerging technologies in the context of neurorehabilitation (for assessment and treatment purposes) has been identified in research studies, which commonly occur in the context of public universities located in specific regions of Brazil (Araújo et al., 2017). As mentioned above, clinicians have difficulty in implementing scientific evidence in their practice. Therefore, evidence on emerging technologies is centered on researchers. Previous studies found that about 60% of the Brazilian clinicians are not involved with research activities (Nascimento et al., 2020; Silva et al., 2015) and access to technologies is restricted to patients who volunteer for research studies. Therefore, these technologies are not commonly available or used routinely by the majority of the professionals and patients in the area of neurorehabilitation. It is possible that some exceptions occur within the context of the clinical practice of some few professionals, who acquired expertise within their own structures, so that their patients could benefit from it. Unfortunately, this is really an exception and cannot be readily identified across this large country.

EMERGING TECHNOLOGIES IN THE CONTEXT OF NEUROREHABILITATION IN BRAZIL

In this section, critical analysis related to the use of emerging technologies in the context of neurorehabilitation will be presented, taking into account two perspectives; first, from the research perspective, because some individuals with neurological conditions may have access to studies that have been carried out with the purpose of developing or investigating the effects of these technologies, and, second, from the perspective of public health, because the majority of the Brazilian population depends on public healthcare services.

Figure 4.2 shows the distribution of research related to the use of emerging technologies in the context of neurorehabilitation, as well as their availability in the context of the public health scenario in Brazil. It is possible to identify that the use of these technologies is concentrated in the southeast of the country, which is a region with higher socioeconomic development. This shows that the other regions of the country lack the provision of these technologies.

We will present some of the most-discussed emerging technologies in the world, that are currently used in Brazil: artificial intelligence, Internet of Medical Things (IoMT), virtual reality, robotics, and telehealth.

ARTIFICIAL INTELLIGENCE

Artificial intelligence involves technologies that have the ability to mimic human intelligence, such as visual perception, speech recognition, and language translation. Emerging technology systems usually have the capacity to learn or adapt to new experiences or stimuli. The concept of artificial intelligence encompasses several

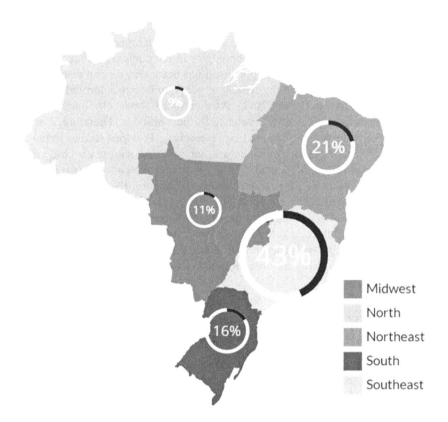

FIGURE 4.2   Regional distribution of the use of emerging technologies in neurorehabilitation in Brazil, in both research and public healthcare.

other concepts, such as machine learning and deep learning (Great Britain, 2017; House of Lords, 2018; Pedersen et al., 2020).

Machine learning is a type of artificial intelligence that allows for learning and improvement with experience, without being explicitly programmed, and machine learning algorithms can make predictions or solve problems when provided with sufficient data (Great Britain, 2017; House of Lords, 2018; Pedersen et al., 2020). Deep learning is a more recent artificial intelligence neural network model, that uses many layers of artificial neurons to solve difficult problems, such as classifying images and sounds (House of Lords, 2018; Pedersen et al., 2020). Artificial intelligence added to clinical data makes it possible to develop complex, comprehensive, and accurate algorithms (Pedersen et al., 2020) and has been used to support clinical decision-making through the development of diagnostic and prognostic models (Magrabi et al., 2019; Pedersen et al., 2020).

In Brazil, investments have been made to expand the use of artificial intelligence. Recently, a Center for Artificial Intelligence (C4AI) was opened in São Paulo, in the southeastern region, through a partnership between the Brazilian Ministry of

Science, Technology and Innovation and a public research institution (Pinhanez, 2020). The Center will initially focus on five great AI challenges, related to healthcare, environment, food production networks, future of the work, and the development of Natural Language Processing technology in Portuguese (Pinhanez, 2020). It is worth noting that one of the lines of research is related to neurorehabilitation (stroke modeling to improve diagnosis, treatment, and rehabilitation) and another focuses on artificial intelligence in emerging countries (Artificial intelligence in emerging countries: public policies and future of work) (Pinhanez, 2020). The goal is to build another seven centers. There are still no available data related to the use of artificial intelligence in the context of neurorehabilitation (FAPESP, 2020). However, it is important to emphasize that advancement in the context of health has been carried out.

Another advance regarding the use of artificial intelligence in the context of health is related to the creation of the Brazilian Strategy for Artificial Intelligence (MCTI, 2021), but the guidelines, published in 2021, only considered the use of artificial intelligence in the context of medicine: teaching medicine, automated exams, pathological analyses, microcellular images and microscopes, and the use of drones for drug delivery (MCTI, 2021). This may explain why research in the context of neurorehabilitation is lacking. On the other hand, areas with the greatest potential growth linked to diagnoses and health treatments were identified (MCTI, 2021). This indicates that possible advances will occur in the future.

Despite seeming to be a relatively new topic, the use of artificial intelligence in neurorehabilitation has been studied in Brazil. For example, Muniz et al. (2010) compared logistic regression (LR), probabilistic neural network (PNN), and support vector machine (SVM) classifiers for discriminating between individuals with and without Parkinson's disease in assessing the effects of brain stimulation of the subthalamic nucleus (DBS-STN) on ground reaction force (GRF) with and without medication. The three models showed high performance indices to classify GRF patterns of normal and untreated individuals with Parkinson's disease. However, using the bootstrap approach, the PNN performed better. In addition, when evaluating the effects of treatments, the PNN was more reliable in discriminating between individuals with and without Parkinson's disease.

Considering the few identified findings, it is possible to conclude that artificial intelligence is still not a technology commonly used in the context of neurorehabilitation in Brazil. In addition to the challenges mentioned in the previous section, the following challenges may also be associated with the limited use of artificial intelligence in the context of neurorehabilitation:

- Thinking that more and more humans will be dependent on machines, and that they will gradually be replaced by them, which would result in unemployment and mechanical stagnation (Meirelles, 2022).
- Concerns related to privacy and data security of the involved people (patients, health professionals, institutions) in the use of this technology (Bernardo, 2020).

- Commitment of the therapist–patient relationship, when the professional is replaced by machines, which can result in a lack of empathy (Bernardo, 2020).
- Lack of professional knowledge (NH, 2022).
- Little investment in research and available resources (Teletime, 2020).

THE INTERNET OF MEDICAL THINGS (IoMT)

The Internet of Things (IoT) is an interconnected network by which physical things, machines, and individuals are connected and can communicate and exchange data through the internet. The Internet of Medical Things (IoMT) is the application of the IoT to healthcare and medicine (Alsubaei et al., 2018; Dwivedi et al., 2022; Hameed et al., 2021). This comprises a collection of medical devices/systems and applications (apps) which are connected through heterogeneous networks (Alsubaei et al., 2018). IoMT technologies can be used to aid diagnosis and evaluation, manage health conditions, monitor medication administration remotely, improve treatment, and deliver telehealth and telerehabilitation (Alsubaei et al., 2018). There are several types of medical devices used in the IoMT system, such as implantable (deep brain implants), wearable (fitness devices), environmental (elderly monitoring devices in smart homes) and stationary devices (medical image processing devices of magnetic resonance imaging and computerized tomography-scan) (Hameed et al., 2021). These medical devices/systems and sensors are responsible for capturing specific data, such as body pressure and blood glucose level, and data are transferred in real time, accumulated, and processed on personal servers located near or far from the patient's body (Hameed et al., 2021).

Data on the use of the IoMT in the context of neurorehabilitation in the public health scenario have not yet been reported. On the other hand, in the scenario of research, data were identified. Various IoMTs can be used to monitor individuals undergoing rehabilitation, such as smartphone apps and watches for monitoring heart rate and physical activity levels. In Brazil, several studies have investigated the use of these apps for monitoring individuals with neurological conditions. Investigations on measurement properties of mobile health (mHealth) devices (Google Fit, Health, STEPZ, Pacer, and Fitbit Inc®) found that they are valid to measure the number of steps undertaken by patients suffering from a stroke or Parkinson's disease (Lana et al., 2021; Costa et al., 2020). The measurement properties of the GT3X® ActiGraph accelerometer and Google Fit® smartphone apps were evaluated in individuals who have had a stroke (Faria et al., 2019; Polese et al., 2019). The results showed that the Google Fit® smartphone app showed adequate measuring properties to estimate energy expenditure in individuals who have suffered a stroke (Polese et al., 2019). However, the GT3X® ActiGraph accelerometer tended to underestimate the data and did not appear to be valid for estimating stepping activity in patients recovering from a stroke (Polese et al., 2019). Both smartphone apps and mHealth devices have advantages, as they are simple, easy to use and access, and relatively inexpensive. In addition, they can be easily used by researchers, clinicians, and family members.

Some smartphone apps are also used to guide the execution of exercises and to facilitate repetitive exercise practice. For example, the REPS Recovery Exercises app features two exercise programs for individuals who have suffered from a stroke. One of the programs involves repetitive practice of daily tasks, whereas the other focuses on repetitive practice of arm movements. The practice of exercises can be viewed by video. However, this app does not have a Portuguese translation, which is a challenge for its use in clinical practice in Brazil. In addition, these resources lack studies that support their use in the context of neurorehabilitation in Brazil.

There are also several rehabilitation websites and apps which are available in the Brazilian-Portuguese language. For example, the website Physiotherapy Exercises for people with injuries and disabilities (www.physiotherapyexercises .com) is freely available and allows for the creation of several exercise programs for patients with neurological conditions, providing demonstrations through photographs and videos. In addition, it features explanatory text for patients and therapists. Another example is the Einstein physiotherapy app, which is also free and allows for clinicians to film custom exercises to guide and monitor patients to perform home-based exercises. Finally, some apps provide a playful platform, such as the FisioAdventure app. This is a free app, in which the individual chooses the theme, which will have different exercise sequences and each series is a different story, all guided by audio.

Considering these findings, it is possible to conclude that the use of the IoMT in the context of neurorehabilitation has mainly focused on the use of apps on mobile devices, since they are the most accessible resources among those available for using the IoMT. However, it is not possible to confirm whether these resources are used within clinical practice, considering that they have mainly been used in research settings. In addition to the challenges mentioned in the previous section, there are several barriers related to the limited use of the IoMT, among which two stand out: difficulty in guaranteeing the protection of patient data and the high cost for private and public hospitals to build a technological system (Vieira & Ponte Junior, 2018).

In 2019, the National Plan for the IoT was created in Brazil by the Ministry of Science, Technology and Innovations with the purpose of implementing and developing the IoT in the country based on free competition and free circulation of data, in compliance with information and data security guidelines and personal data protection (Brasil, 2019, 2021). The strategic actions involved the creation of platforms for innovation in health, the recognition of health informatics as an area of knowledge on the part of education bodies, and the offer of scholarships for masters, doctoral, post-doctoral students and research studies, and the creation of post-graduate courses that bring together people with a background in mathematics, physics, computer science, or engineering with people in health fields (Brasil, 2019, 2021). No information related to the implementation and results of these strategic actions is available, but it is important to consider that the creation of a national plan is an advance toward the use of this technology.

Virtual Reality

Virtual reality is a collection of technologies that allow for the interaction of a human–machine interface through a computer-generated synthetic environment (Gigante, 1993; Kardong-Edgren et al., 2019; Narváez et al., 2017; Perez-Marcos, 2018). Virtual reality is an immersive and multisensory (visual, auditory, tactile, sensory, kinesthetic, and proprioceptive) experience that allows for the individual to recreate the sensation provided by artificial sensory feedback, in environments that appear to be and feel similar to real-world activities and events (Narváez et al., 2017; Perez-Marcos, 2018). Several devices/systems allow for the simulation of real clinical scenarios for patients and health professionals (Perez-Marcos, 2018). These experiences can be achieved by using simple devices/systems (e.g., computers or cellphone cameras) or more complex ones, such as viewing helmets (Perez-Marcos, 2018).

In Brazil, several research groups have investigated the effects of virtual reality in the context of neurorehabilitation. In some places, it has also been used within clinical contexts through devices, such as Nintendo Wii and Xbox for both assessment and interventions. Moura et al. (2021) used a virtual functional mobility test (VFMT) as a clinical assessment tool for functional mobility of individuals with Parkinson's disease, during simple and complex tasks. This was developed with Unity™ software, and can be exported to multiple platforms, such as Xbox, Android, iOS, Windows, Linux, Mac, and PlayStation. A Microsoft Kinect V2 (Microsoft Corp., Redmond, WA, USA) sensor was used to detect movements. To perform the test, individuals control an avatar to move through a simulated house and interact with various everyday objects. The virtual functional mobility test showed adequate feasibility, sensitivity, and inter-rater reliability.

Several studies carried out in Brazil investigated the effects of interventions using virtual reality therapy. The effects of virtual reality training on walking distance and physical fitness were investigated on individuals with Parkinson's disease, using the Kinect Xbox 360™ (Microsoft Corporation, Redmond, WA, USA) motion sensor. Gait training via virtual reality was as effective as treadmill training in improving walking distance and temporal gait variables (Melo et al., 2018). In addition, Ferraz et al. (2018) found that the effects of virtual reality training, functional training, and stationary bicycle exercise on walking capacity, ability to stand up and sit, and functionality were similar to those of conventional physical therapy. Finally, Mendes et al. (2012) investigated learning, retention, and transfer of performance improvements after virtual reality training, using the platform-based video game Nintendo WiiFit™. The results showed that individuals with Parkinson's disease showed no deficits in learning or retention on seven of the ten games, when compared with healthy individuals. In addition, they were able to transfer motor ability trained on the games to a similar untrained task.

Virtual reality has also been applied to individuals suffering from a stroke. Miranda et al. (2019) investigated the transfer of gains obtained after balance training with Nintendo WiiFit™ to an untrained task with similar balance demands. The results showed that the experimental group showed statistically significant improvements in scores in all five games after training. However, there was no transfer of the gains to an untrained task with similar balance demands. According to the authors, the possible reasons for these results were that the amount of training was insufficient to promote a consolidation of learning, and the difference in training and evaluation contexts. Junior et al. (2019) compared a program combining virtual reality games and proprioceptive neuromuscular facilitation with the stand-alone techniques on sensorimotor performance. A Nintendo Wii device was also used, and four electronic games were included in the protocol: Balance Bubble Plus, Rhythm Parade, Tennis, and Box. The results showed that the effects of combining virtual reality and proprioceptive neuromuscular facilitation were similar to those obtained with stand-alone techniques.

The effects of interventions using virtual reality were also evaluated in children with neurological conditions. Lazzari et al. (2015) investigated the effects of a single session of transcranial direct current stimulation combined with virtual reality training on balance of children with cerebral palsy. Virtual reality mobility training using an XBOX 360™ with a Kinect™ and the Your Shape Fitness Evolved 2012™ game was conducted with simultaneous transcranial direct current stimulation (active or placebo). Increases in body sway velocity were found in the group that received active transcranial direct current stimulation. These results indicate the potential of this form of stimulation in the treatment of individuals with neurological conditions, as well as causing changes in the excitability of the cortex. Grecco et al. (2015) investigated the effects of anodal transcranial direct current stimulation, compared with sham, during gait training with virtual reality on gait patterns, gross motor function, functional performance, and cortex excitability in children with spastic cerebral palsy, also using the XBOX 360™ with a Kinect™ device. The results showed increases in cortex excitability and positive effects in spatiotemporal gait variables (velocity and cadence), gross motor function, and mobility.

Virtual reality has also been used for neurorehabilitation in the public health scenario. Virtual reality glasses are being used in two public hospitals in the south of Brazil (Andrade, 2022; Paranashop, 2022). Other technologies related to virtual reality have not yet been identified in the public health scenario. Currently, virtual reality glasses at low cost are available on the market (cost about US$10) and can be used with a cell phone (not requiring other, more expensive resources, such as computers) (Queiroz, 2022). Therefore, it is a technology that has the potential to be used in various scenarios.

Unlike other mentioned technologies, virtual reality has frequently been researched in the context of neurorehabilitation in Brazil. However, it is still not widely used in public health settings. In addition to the challenges mentioned in the previous section, related to the use of emerging technologies in Brazil, other challenges may also be associated with the limited use of virtual reality in the context of neurorehabilitation. Bezerra & Souza (2018) reported the following barriers related to the use of virtual reality with children and adolescents identified by occupational therapists: the adaptation of the devices, the implementation of the resource in rehabilitation institutions, the high cost, and the difficulty in choosing appropriate games. These barriers are also possibly identified in clinical practice of other health professionals involved in the rehabilitation of individuals with neurological conditions.

ROBOTICS

Rehabilitation with robots, involving human–machine interactions and programmed robotic devices, may assist in the process of evaluating, treating, and monitoring of individuals (Narváez et al., 2017). These devices can support and enhance the productivity and effectiveness of the clinicians and facilitate rehabilitation of individuals with disabilities (Langhorne et al., 2011). In addition, they allow for the professionals to control various parameters during rehabilitation sessions, such as speed, torque, and range of motion (Krebs & Volpe, 2013). Robotic devices are jointly developed and managed by patients, engineers, designers, and therapists (Langhorne et al., 2011).

In Brazil, several research groups have investigated the use of robotics in the context of neurorehabilitation. Terranova et al. (2021) compared the effects of Robot-Assisted Therapy and Constraint-Induced Movement Therapy with chronic stroke patients. The two therapies showed similar results and improvements were found in overall upper limb function, motor recovery, motor function, and activities of daily living, regardless of the interventions. Santos & Siqueira (2018) investigated the effects of a robotics-assisted gait training with a robot-driven exoskeleton orthosis equipment Lokomat® 5.0, compared with therapy-assisted training, on balance, coordination, and independence in activities of daily living in individuals with stroke with ataxia. The two groups showed similar improvements in balance and functional independence in daily living activities, and in general ataxia sequelae symptoms.

The effects of robotics on lower limb motor function were investigated in individuals suffering from a stroke. Silva-Couto et al. (2019) investigated whether a single robotic therapy session would promote short-term ankle adaptations that could influence ankle torque stability and walking speed in 13 individuals with chronic stroke and 13 healthy controls, matched by age and gender. Individuals with stroke showed gains in short-term performance in torque stability, especially during ankle dorsiflexion, after a single robotic therapy session. Robotic therapy did not influence walking speed, although low to moderate correlations were observed between torque stability variables and walking speed.

Emerging technology involving robotics has also been used in the context of neurorehabilitation in the public health scenario in Brazil. Examples of these technologies, their approximate cost, and where they can be accessed are shown in Table 4.2. They include cycle ergometers for the upper and lower limbs and body support that allows the maintenance of the individual's orthostatic position associated with functional electrical stimulation on the lower limb muscles. All these technologies can be used with individuals who have disabilities due to neurological conditions. For example, the Ergys and Erigon technologies allow for individuals with spinal cord injuries to perform exercises with the lower limbs, contributing to improvements in lower limb function and cardiorespiratory fitness (NeuroRehab Directory, 2022a; Portal do Governo do Estado de São Paulo, 2014; Setor Saúde, 2018). The Armeo

TABLE 4.2

Description of the Use of Emerging Technologies Involving Robotics in the Public Health Scenario in Brazil

Technology	Description	Approximate Cost	Location
Erigo® (basic and pro) (NeuroRehab Directory, 2022a; Setor Saúde, 2018)	It consists of an orthostatic plate for robotic rehabilitation of the lower limbs combined with computerized functional electrical stimulation for specific lower limb muscles. It involves gradual verticalization of the load with cyclic movement of the leg.	$10,000	Public hospital: Santa Casa de Misericórdia. Porto Alegre, Rio Grande do Sul.
ERGYS (Portal do Governo do Estado de São Paulo, 2014)	It consists of a lower limb cycle ergometer associated with functional electrical stimulation.	Unknown	Public hospital: Instituto de Reabilitação Lucy Montoro, São Paulo.
MobiTronics (Portal Hospitais Brasil, 2021)	It is a lower limb cycle ergometer associated with functional electrical stimulation, to be used at the bedside.	Unknown	Public hospital: Santa Casa de Misericórdia, Porto Alegre, Rio Grande do Sul.
InMotion Arm (NeuroRehab Directory, 2022b; Ágil Representação, 2017; Ossanai et al., 2019)	It consists of a robotic upper limb exoskeleton that assists in shoulder, elbow, and wrist movements.	$10,000	Public hospital: Instituto de Reabilitação Lucy Montoro, São Paulo. Associação de Assistência à Criança Deficiente (AACD) (provides services via the single health system and private health plans, São Paulo).
Armeo Spring (Ágil Representação, 2017; SMBrasil, 2022; APAE Campo Grande, 2018)	It consists of an upper limb support, that facilitates active shoulder, elbow, wrist, and hand movements during exercises. It is used through the simulation of specific tasks and games.	Unknown	Public hospital: Instituto de Reabilitação Lucy Montoro, São Paulo.

(Continued)

TABLE 4.2 (CONTINUED)

Description of the Use of Emerging Technologies Involving Robotics in the Public Health Scenario in Brazil

Technology	Description	Approximate Cost	Location
Lokomat (MaisGoiás, 2022; Ágil Representação, 2017; AACD, 2020; APAE Campo Grande, 2018)	It consists of a robotic exoskeleton composed of a platform that supports the patient's pelvic girdle and two orthoses for the lower limbs, which allow for the use of the hip and knee joints, facilitating locomotion on the treadmill.	$380,000	Public hospital: Instituto de Reabilitação Lucy Montoro, São Paulo. Public hospital: Centro Estadual de Reabilitação e Readaptação Dr. Henrique Santillo (Crer), Goiás. Public health clinic: Centro Especializado em Reabilitação da APAE de Campo Grande, Mato Grosso do Sul.

Spring and InMotion technologies allow for individuals with traumatic brain injury, multiple sclerosis, cerebral palsy, or stroke to perform exercises with the upper limbs, also contributing to improvements in lower limb function and cardiorespiratory fitness (Ágil Representação, 2017; APAE Campo Grande, 2018; NeuroRehab Directory, 2022b; Ossanai et al., 2019; SMBrasil, 2022).

The costs of these technologies are not clear, but it is important to consider that they were created in developed countries. For instance, the InMotion was developed in the United States and the Lokomat and Armeo Spring in Switzerland. Among the mentioned examples, there are technologies with a cost of approximately US$380,000. Possibly, there are not many models of this technology in Brazil. This shows the need for greater investment in the production of national robotic technologies, which would contribute to improving the supply of health services, as well as to the scientific development of a country such as Brazil. Finally, it is important to consider that some technologies involving robotics are available in public hospitals, which allows for greater accessibility to these resources for individuals who depend on the public health services.

Telehealth and Telerehabilitation

The use of technologies in health care has become more evident with the advancement of telehealth (Narváez et al., 2017). Telehealth is the "use of electronic information and telecommunication technologies to support long-distance clinical health care, patient, and professional health-related education, health administration, and public health." Telehealth is a broad term that encompasses other emerging terms, such as telerehabilitation (Health Resources & Services Administration, 2022). Telerehabilitation is considered to be a growing alternative to delivering rehabilitation services, using information and communication technologies (Brennan et al., 2009). The use of telerehabilitation in Brazil and in several other countries has grown significantly since the beginning of the COVID-19 pandemic, as this was an important strategy to deliver rehabilitation care. It has several advantages, such as the provision of assistance to patients in their own homes or residences, which makes it possible to deliver rehabilitation to people living in geographically distant locations, and reduces transportation time, discomfort, and costs (Hjelm, 2005). This is particularly important in Brazil, which is a country with continental dimensions. These aspects are important in individuals with neurological conditions, who commonly have impairments that prevent them from accessing face-to-face rehabilitation. In addition, telerehabilitation may be more cost-effective than face-to-face rehabilitation (Hjelm, 2005).

Information and communication technologies (ICT) were used to evaluate and monitor individuals remotely in telehealth/telerehabilitation. Telephone calls, mHealth devices, and smartphone apps were used in some studies conducted in Brazil. Camozzato et al. (2011) evaluated the test-retest reliability and diagnostic validity of the Brazilian telephone Mini-Mental State Examination (Braztel-MMSE) in individuals with Alzheimer's disease. The telephone-based version of the MMSE was compared with the original MMSE applied in-person. The telephone-based version showed adequate validity and reliability, when compared with the face-to-face application. In addition, it demonstrated good diagnostic properties for screening for dementia in individuals with Alzheimer's disease. The measurement properties of the Brazilian versions of the telephone-based application of the modified Rankin Scale (mRS), Interview for Cognitive Status (TICS-M), ABILHAND, and ABILOCO were also investigated in individuals with stroke (Avelino et al., 2020; Avelino et al., 2021; Baccaro et al., 2015; Baggio et al., 2014). All showed adequate validity and reliability when administered by telephone

Rehabilitation may also involve health education benefits for the individuals with neurological conditions and their caregivers/family members, as well as for health professionals. Two apps have been identified on Rede Brasil AVC website (RedeBrasilAVC, 2021), which is a non-governmental organization created in 2008 with the goal of improving global care for individuals who have suffered a stroke. This organization has a website that delivers information for both patients and health professionals. In addition, it provides information on two smartphone apps, which are freely available: AVCBrasil and Stroke Riskometer. The AVCBrasil app aims to describe the main symptoms of a stroke, to list good lifestyle habits, and state the location of specialized stroke centers, besides allowing the making of emergency calls through buttons visible on the app screen. The Stroke Riskometer app suggests strategies to improve health and reduce risk factors, in addition to teaching to recognize signs of stroke in order to get help quickly. Even though they lack scientific evidence to confirm their usability and efficacy in improving outcomes, the Brazilian stroke guidelines recommend the use of the Stroke Riskometer in the context of secondary stroke prevention (Ministério da Saúde, 2020).

Related to the prevention scenario, Maniva (2016) developed and validated an educational booklet to prevent stroke recurrence, entitled "Cartilha do AVC: o que é, o que fazer e como prevenir" (in English, "Booklet about stroke: what it is, what to do, and how to prevent it"). It is composed of seven sections: definition of the disease, signs and symptoms, emergency actions, treatment, risk factors, recurrence, and preventive measures. The booklet showed an excellent degree of agreement among specialists and proved to be capable of promoting knowledge, attitude, and adequate practice in individuals who have had a stroke. This type of material is common in Brazil, and has the advantage of being able to be used in both printed and digital formats. In digital format, considering the advancement of the telerehabilitation scenario, the material can be delivered to individuals through social media and e-mail.

Finally, these apps can also be used to assist and improve the routine of clinicians. Martins et al. (2020) investigated the use of the JOIN App smartphone system to expedite decision-making in individuals who had suffered a stroke. JOIN App enables quick sharing of clinical patient data. In the JOIN system, Digital Imaging and Communications in Medicine images are exported by a virtual server located within the hospital network. The anonymized Digital Imaging and Communications in Medicine are sent to the Amazon Cloud Server through the Virtual Private Network, from where the images are transmitted to the app installed on the professionals' smartphones. The JOIN App had various features, such as chat and image viewing function through Digital Imaging and Communications in Medicine system (e.g., computerized tomography and magnetic resonance imaging). The results showed that the app was effective and safe to connect with the stroke team inside and outside the hospital. In addition, there was a significant impact on workflow, as evidenced by a significant reduction in door-to-needle times (Martins et al., 2020).

The Brazilian Telehealth Network Program, implemented by the Ministry of Health (Brasil, 2022), has several objectives, such as: overcoming socioeconomic, cultural, and, above all, geographic barriers, so that health services and information can reach the entire population; greater user satisfaction, higher quality of care, and lower cost for the unified health system; reduced waiting lines; and avoidance of unnecessary displacements of patients and health professionals. In addition, this program has the following fields of activity: teleconsulting, telediagnosis, telemonitoring, teleregulation, and teleeducation. There are 26 Telehealth Centers in Brazil, located in the five regions of the country: Northeast (38%), North (19%), Southeast (19%), South (12%), and Midwest (12%). Although the other technologies were concentrated in the South and Southeast regions, telehealth centers are more concentrated in the Northeast region, due to the greater need for actions aimed at controling the COVID-19 pandemic. However, it is important to emphasize that data related to the use of these nuclei in the neurorehabilitation scenario are not yet available. In addition to the challenges mentioned in the previous sections, related to the use of emerging technologies, other challenges may also be associated with the limited use of telehealth and telerehabilitation in the context of neurorehabilitation in Brazil: difficulty experienced by participants and caregivers in using necessary technologies, such as computers, and difficulty of patients in communicating and expressing themselves (Cabral, 2021).

ETHICAL AND MORAL ASPECTS RELATED TO THE USE OF EMERGING TECHNOLOGIES IN THE CONTEXT OF NEUROREHABILITATION IN BRAZIL

Several ethical and moral aspects, which could compromise the use of emerging technologies in neurorehabilitation worldwide, should be considered. There are factors to be re-considered, in view of the specificity of each location and the ethnic characteristics of the Brazilian population. Brazil is a middle-income country, and a multicultural and ethnically diverse nation that faces several challenges for the implementation of emerging technologies. These cause several ethical concerns and moral difficulties for health professionals. These aspects will be addressed, as follows:

- **Equity in the health system** This is an important ethical aspect that must be considered, especially in the context of a developing country. As previously mentioned, inequalities (educational, economic, social, and cultural) may compromise access to emerging technologies. Equity is essential in any health system. Access to the same therapeutic possibilities must be guaranteed to all individuals, without exclusion, discrimination, or stigmatization. However, the principle of equity could be affected, when not all patients have access to the appropriate technologies (Brall et al., 2019; Gómez-Ramírez et al., 2021). It is essential that these factors are carefully analyzed, when considering the use of these technologies in clinical practice.
- **Digital tools in the absence of a robust health framework** As previously mentioned, Brazil is a country that faces several difficulties in the implementation of technologies. The lack of a robust framework in the health system for the implementation of emerging technologies impacts their accessibility and functioning, in terms of the ability of both professionals and patients to use them and the privacy and safety concerns regarding patients (Brall et al., 2019). This scenario makes implementation more challenging and adds ethical and moral concerns for practitioners.
- **Implementation of evidence-based practice** There is evidence for the use of emerging technologies to enhance assessment and improve various outcomes in neurorehabilitation (Brall et al., 2019; Gómez-Ramírez et al., 2021). However, in Brazil, there are several difficulties in implementing these technologies within clinical contexts, which centralizes their use in research groups. In addition, the main neurorehabilitation guidelines, which are available in Portuguese, do not address the use of these technologies. Thus, clinicians find it difficult to use the best scientific evidence in their clinical practice when it involves the use of emerging technologies.

- **Dignity and autonomy** When using emerging technologies in clinical practice, health professionals must reflect on and critically evaluate whether the patient's dignity is being respected, such as when using telehealth/telerehabilitation with vulnerable people, such as those with cognitive impairments (Brall et al., 2019; Gómez-Ramírez et al., 2021). This is particularly important in Brazil, since there are many patients in situations of social vulnerability. In addition, patient's autonomy must be respected. The patients and their families should be consulted and provide consent regarding the use of emerging technologies. Brazil is one of the most multicultural and ethnically diverse nations in the world. Therefore, the autonomy and diversity of patients' preferences must be taken into account when any rehabilitation strategies, including the use of emerging technologies, are planned.
- **Social impact** When implementing a new technology in the health system, its social impact must be evaluated. One must critically assess the technology's advantages over existing interventions, cost-effectiveness, and impact for patients and the health system (Brall et al., 2019; Gómez-Ramírez et al., 2021). This is particularly important in the Brazilian context, where resources for the implementation of new health technologies are limited
- **Privacy and security** One of the main problems with emerging technologies is the fragility of security (Hameed et al., 2021). Some devices and sensors do not support security features, such as encrypted messages. Furthermore, the systems are heterogeneous, meaning that not all security measures are applied to all devices. The greater the use of health technologies and the more devices connected to the network, the greater the risk of attacks, privacy breaches, and data leakage. Moreover, cybersecurity incidents on medical devices can have immeasurable consequences for the health and safety of patients. Therefore, security and privacy become a concern. Some metrics are essential for IoMT systems, such as confidentiality, integrity, availability, privacy, and non-repudiation. Any event that threatens any of these features is considered an attack (Hameed et al., 2021). In Brazil, the General Personal Data Protection Law (Law n° 13.709/2018) regulates rules for the use of personal data. In addition, consequences can be foreseen for cases of inappropriate treatment and data leakage. Therefore, both clinicians and researchers must be careful when choosing a device or sensor, so that the security and privacy of their patients are maintained.

FUTURE PROJECTIONS ON THE USE OF EMERGING TECHNOLOGIES IN NEUROREHABILITATION IN BRAZIL

With the COVID-19 pandemic situation, there were important technological advances and increases in the use of technologies in clinical practice, especially in the context of telehealth and telerehabilitation. This is projected to increase even more in the coming years but, for that to occur, many challenges need to be faced. A critical analysis of the use of emerging technologies in neurorehabilitation in Brazil was carried out in this chapter. As presented, in the current scenario, technologies are centralized in research centers and their use is still quite limited within clinical contexts.

It is important to point out that emerging technologies should not be considered the all-encompassing panacea to healthcare, as they can fail to meet the needs of individuals and/or perhaps those "hard-to-reach" groups/cultures. The use of technologies in clinical practice should be considered with caution, taking into account cost, accessibility, feasibility, clinical utility, and patients' preferences. In addition, it is urgently necessary that the main Brazilian guidelines address the use of emerging technologies in neurorehabilitation.

Currently, several factors limit the use of technologies in clinical practice, with their costs being important barriers to their adoption. In this context, efforts to conduct research aimed at developing low-cost technologies must be carried out. For example, a research group in the northeast of Brazil is developing an orthosis that will help upper limb movements of individuals with amyotrophic lateral sclerosis (UFRN, 2020). The researchers aim at developing an orthosis that is as inexpensive as possible (between US\$600 to US\$750) to facilitate access (UFRN, 2020). Other Brazilian research groups planning to develop or investigate the use of emerging technologies in the context of neurorehabilitation need to take costs into account.

The development of teaching materials with language accessible for both clinicians and patients is very important and can increase the feasibility of its use in clinical practice. A research group in the northeast of Brazil that aims at promoting technological innovation in health (LAIS, 2021) has been working along three lines: management, assistance, and education. This group has published physical and digital articles and books, registered software, and patented apps, as well as providing community-based activities. Other research groups aiming to develop teaching materials in the context of neurorehabilitation need to take this aspect into account.

Finally, efforts have been made to address the use of emerging technologies in guidelines published in the Brazilian-Portuguese language. This is proven by the clinical protocols and therapeutic guidelines for individuals with neurological conditions prepared by CONITEC. The guideline developed for individuals with multiple sclerosis emphasizes the importance of a multidisciplinary team approach, but focuses on guiding the use of medications, rather than other essential technologies for the follow-up of these individuals (Comissão nacional de incorporação de tecnologias no Sistema Único de Saúde, 2021). Similarly, the guidelines developed for individuals with Parkinson's disease also focus on recommending the use of drugs without any recommendation regarding important emerging technologies in neurorehabilitation (Comissão nacional de incorporação de tecnologias no Sistema Único de Saúde, 2021).

If, on the one hand, there are a number of barriers that make the use of emerging technologies still incipient in Brazil, on the other hand, it is important to consider that these technologies do not apply to all contexts of healthcare needs. The implementation of any treatment strategy must meet the needs of patients and be used considering their preferences, one of the pillars of evidence-based practice, and following what is advocated by patient-centered practice. Therefore, it is necessary to consider that emerging technologies, like any other treatment strategies, may not meet the needs and/or preferences of all patients. This may be particularly common in Brazil, a country of continental proportions and characteristics, with great diversity of ethnicities and cultures. Brazil is one of the largest multicultural countries and a nation of great ethnic diversity. Therefore, even if the current barriers that hinder the use of emerging technologies in Brazil are overcome, emerging technologies, like any other treatment strategy, must be prescribed according to client preferences and client-centered practice. It is up to the health professional to carefully conduct their assessment, following the basic assumptions of professional performance, such as evidence-based practice and client-centered practice, and considering not only the "virtues of emerging technologies" but also provide a view to "critically examine its application to assist service users/patients now and in future years."

REFERENCES

AACD. *Pacientes imobilizados poderão voltar a andar com ajuda de armadura robótica no CRER.* 2020. Available in: https://aacd.org.br/noticias/pacientes-imobilizados-poderao -voltar-a-andar-com-ajuda-de-armadura-robotica-no-crer. Access date: July 28, 2022.

Ágil Representação. *Alta tecnologia em reabilitação robótica. Robótica na reabilitação.* 2017. Available in: https://www.agilrepresentacao.com.br/single-post/2017/06/22/rob %C3%B3tica-na-reabilita%C3%A7%C3%A3o-1. Access date: July 28, 2022.

Alsubaei F, Abuhussein A, Shiva S. A framework for ranking IoMT solutions based on measuring security and privacy. *Adv Intell Syst Comput.* 2018;880:205–224. doi:10.1007/978-3-030-02686-8_17.

Andrade A. *Viva Bem. Longevidade. Práticas e atitudes para uma vida longa e saudável. Hospital de Curitiba usa realidade virtual para tratar idosos internados.* 2022. Available in: https://www.uol.com.br/vivabem/noticias/redacao/2022/01/21/hospital -de-curitiba-usa-realidade-virtual-para-tratar-idosos-internados.htm. Access date: July 28, 2022.

Anwar F, Shamim A. Barriers in adoption of health information technology in developing societies. *Int J Adv Comput Sci Appl.* 2011;2(8):40–45.

APAE Campo Grande. Notícias. *CER/APAE faz lançamento do primeiro Laboratório de Robótica do Centro-Oeste.* 2018. Available in: https://apaecg.org.br/cer-apae-faz-lan-camento-do-primeiro-laboratorio-de-robotica-do-centro-oeste/. Access date: July 28, 2022.

Araújo DV, Distrutti MSC, Elias FTS. Priorização de tecnologias em saúde: o caso brasileiro. *J Bras Econ Saúde.* 2017;9(Suppl.1):4–40. doi:10.21115/JBES.v9.suppl1.4-40.

Avelino PR, Menezes KKP, Nascimento LR, et al. Validation of the telephone-based applica-tion of the ABILHAND for assessment of manual ability after stroke. *J Neurol Phys Ther.* 2020;44(4):256–260. doi:10.1097/NPT.0000000000000326.

Avelino PR, Nascimento LR, Menezes KKP, et al. Validation of the telephone-based assessment of locomotion ability after stroke. *Int J Rehabil Res.* 2021;44(1):88–91. doi:10.1097/MRR.0000000000000447.

Baccaro A, Segre A, Wang YP, et al. Validation of the Brazilian-Portuguese version of the modified telephone interview for cognitive status among stroke patients. *Geriatr Gerontol Int.* 2015;15(9):1118–1126. doi:10.1111/ggi.12409.

Baggio JA, Santos-Pontelli TE, Cougo-Pinto PT, et al. Validation of a structured interview for telephone assessment of the modified Rankin Scale in Brazilian stroke patients. *Cerebrovasc Dis.* 2014;38(4):297–301. doi:10.1159/000367646.

Bernardo A. *Inteligência artificial: Ela está no meio de nós.* 2020. Available in: https://saude
.abril.com.br/medicina/inteligencia-artificial-ela-esta-no-meio-de-nos/. Access date:
July 28, 2022.

Bezerra TF, Souza VLV. O uso da realidade virtual como um recurso terapêutico ocupa-
cional na reabilitação neurológica infanto-juvenil. *Rev Interinst Bras Ter Ocup.*
2018;2(2):272–291.

Brall C, Schröder-Bäck P, Maeckelberghe E. Ethical aspects of digital health from a justice
point of view. *Eur J Public Health.* 2019;29(Supplement 3):18–22. doi:10.1093/eurpub/
ckz167.

Brasil. *Decreto nº 9.854, de 25 de junho de 2019. Institui o plano nacional de internet das
coisas e dispõe sobre a câmara de gestão e acompanhamento do desenvolvimento de
sistemas de comunicação máquina a máquina e internet das coisas.* 2019. Available in:
https://www.planalto.gov.br/ccivil_03/_ato2019-2022/2019/decreto/d9854.htm. Access
date: July 28, 2022.

Brasil. *Governo digital. Estratégias e políticas digitais. Plano nacional de internet das coisas.* 2021. Available in: https://www.gov.br/governodigital/pt-br/estrategias-e-politicas
-digitais/plano-nacional-de-internet-das-coisas. Access date: July 28, 2022.

Brasil. *Ministério da Saúde. Diretrizes de atenção à reabilitação da pessoa com acidente
vascular cerebral.* Brasília: Ministério da Saúde, 2013.

Brasil. *Ministério da Saúde. Portaria conjunta nº 10, de 31 de outubro de 2017. Aprova
o protocolo clínico e diretrizes terapêuticas da Doença de Parkinson.* Available in:
https://www.gov.br/saude/pt-br/assuntos/protocolos-clinicos-e-diretrizes-terapeuticas
-pcdt/arquivos/2022/portaria-conjunta-no-10-2017-pcdt-doenca-de-parkinson.pdf.
Access date: July 28, 2022.

Brasil. *Ministério da Saúde. Portaria conjunta nº 13, de 13 de agosto de 2020. Aprova o protocolo clínico e diretrizes terapêuticas da esclerose lateral amiotrófica.* Available in:
https://www.gov.br/saude/pt-br/assuntos/protocolos-clinicos-e-diretrizes-terapeuticas
-pcdt/arquivos/2020/portaria_conjunta_pcdt_ela.pdf. Access date: July 28, 2022.

Brasil. *Programa Telessaúde.* Available in: https://www.gov.br/saude/pt-br/acesso-a-infor-
macao/acoes-e-programas/programa-telessaude. Access date: July 28, 2022.

Brennan DM, Mawson S, Brownsell S. Telerehabilitation: Enabling the remote delivery
of healthcare, rehabilitation, and self management. *Stud Health Technol Inform.*
2009;145:231–248.

Brito SAF, Aguiar LT, Garcia LN, et al. Cardiopulmonary exercise testing and aerobic treadmill training after stroke: Feasibility of a controlled trial. *J Stroke Cerebrovasc Dis.*
2020;29(7):104854.

Brito SAF, Scianni AA, Peniche PDC, et al. Measurement properties of outcome measures used in neurological telerehabilitation: A systematic review protocol. *PloS One.*
2022;17(3):e0265841.

Cabral FC. *Telemedicina na prática: Principais barreiras e facilitadores.* 2021. Available
in: https://www.hospitalmoinhos.org.br/atrion/pt_BR/noticias/telemedicina-na-pratica
-principais-barreiras-e-facilitadores. Access date: July 28, 2022.

Camozzato AL, Kochhann R, Godinho C, et al. Validation of a telephone screening test for
Alzheimer's disease. *Aging Neuropsychol Cogn.* 2011;18(2):180–194. doi:10.1080/138
25585.2010.521814.

Chor D, Lima CR. Aspectos epidemiológicos das desigualdades raciais em saúde no Brasil
[Epidemiologic aspects of racial inequalities in health in Brazil]. *Cad Saude Publica.*
2005;21(5):1586–1594. doi:10.1590/s0102-311x2005000500033.

Comissão nacional de incorporação de tecnologias no Sistema Único de Saúde. *A comissão.*
2021. Available in: http://conitec.gov.br/entenda-a-conitec-2. Access date: July 28,
2022.

Conselho Nacional de Saúde. *Resolução nº 671, de 05 de abril de 2022*. 2022.
Costa PHV, de Jesus TPD, Winstein C, et al. An investigation into the validity and reliability of mHealth devices for counting steps in chronic stroke survivors. *Clin Rehabil.* 2020;34(3):394–403. doi:10.1177/0269215519895796.
Dwivedi R, Mehrotra D, Chandra S. Potential of Internet of Medical Things (IoMT) applications in building a smart healthcare system: A systematic review. *J Oral Biol Craniofac Res.* 2022;12(2):302–318. doi:10.1016/j.jobcr.2021.11.010.
FAPESP. *Chamada de Propostas FAPESP – MCTIC - CGI.BR para centros de pesquisas aplicadas em inteligência artificial.* 2020. Available in: https://fapesp.br/13896/chamada-de-propostas-fapesp-mctic-cgibr-para-centros-de-pesquisas-aplicadas-em-inteligencia-artificial. Access date: July 28, 2022.
Faria GS, Polese JC, Ribeiro-Samora GA, et al. Validity of the accelerometer and smartphone application in estimating energy expenditure in individuals with chronic stroke. *Braz J Phys Ther.* 2019;23(3):236–243. doi:10.1016/j.bjpt.2018.08.003.
Ferraz DD, Trippo KV, Duarte GP. The effects of functional training, bicycle exercise, and exergaming on walking capacity of elderly patients with Parkinson disease: A pilot randomized controlled single-blinded trial. *Arch Phys Med Rehabil.* 2018;99(5):826–833. doi:10.1016/j.apmr.2017.12.014.
Gigante MA. 1 - Virtual reality: Definitions, history and applications. *Virtual Reality Systems.* 1993;3–14. doi:10.1016/B978-0-12-227748-1.50009-3.
Gómez-Ramírez O, Iyamu I, Ablona A, et al. On the imperative of thinking through the ethical, health equity, and social justice possibilities and limits of digital technologies in public health. *Can J Public Health.* 2021;112(3):412–416.
Great Britain. *Department for business, energy and industrial strategy. Industrial strategy: Building a Britain fit for the future.* 2017. Available in: https://assets.publishing.service.gov.uk/government/uploads/system/uploads/attachment_data/file/664563/industrial-strategy-white-paper-web-ready-version.pdf. Access date: July 28, 2022.
Grecco LAC, Duarte NAC, Mendonça ME, et al. Effects of anodal transcranial direct current stimulation combined with virtual reality for improving gait in children with spastic diparetic cerebral palsy: A pilot, randomized, controlled, double-blind, clinical trial. *Clin Rehabil.* 2015;29(12):1212–1223. doi:10.1177/0269215514566997.
Hameed SS, Hassan WH, Abdul Latiff L, et al. A systematic review of security and privacy issues in the internet of medical things; the role of machine learning approaches. *Peer J Comput Sci.* 2021;7:e414. doi:10.7717/peerj-cs.414.
Hasenbalg CA, Silva NV. Notes on racial and political inequality in Brazil. In: Hanchard M, editor. *Racial politics in contemporary Brazil.* Durham/London: Duke University Press; 1999:154–178.
Health Resources & Services Administration. *What is telehealth? Health resources & services administration.* 2022. Available in: https://www.hrsa.gov/rural-health/telehealth/what-is-telehealth. Access date: July 28, 2022.
Hjelm NM. Benefits and drawbacks of telemedicine. *J Telemed Telecare.* 2005;11(2):60–70. doi:10.1258/1357633053499886.
House of Lords. *AI in the UK: Ready, willing and able?: Report of session 2017-19.* 2018. Available in: https://publications.parliament.uk/pa/ld201719/ldselect/ldai/100/100.pdf. *Access date:* Access date: July 28, 2022.
Instituto Brasileiro de Geografia e Estatística (IBGE). *Pesquisa nacional por amostra de domicílios contínua. Acesso à internet e à televisão e posse de telefone móvel celular para uso pessoal.* 2019. Available in: https://biblioteca.ibge.gov.br/visualizacao/livros/liv101794_informativo.pdf. Access date: July 28, 2022.

Instituto Brasileiro de Geografia e Estatística (IBGE). *Pesquisa nacional por amostra de domicílios contínua, 2012–2019 (acumulado de primeiras visitas), a partir de 2020 (acumulado de quintas visitas).* Available in: https://sidra.ibge.gov.br/pesquisa/pnadca/ tabelas. Access date: July 28, 2022.

Junior VADS, Santos MDS, Ribeiro NMDS, et al. Combining proprioceptive neuromuscular facilitation and virtual reality for improving sensorimotor function in stroke survivors: A randomized clinical trial. *J Cent Nerv Syst Dis.* 2019;11. doi:10.1177/1179573519863826.

Kardong-Edgren SS, Farra SL, Alinier G, et al. A call to unify definitions of virtual reality. *Clin Simul.* 2019;31:28–34. doi:10.1016/j.ecns.2019.02.006.

Krebs HI, Volpe BT. Rehabilitation robotics. *Handb Clin Neurol.* 2013;110:283–294. doi:10.1016/B978-0-444-52901-5.00023-X.

Laboratório de Inovação Tecnológica em Saúde (LAIS). *Universidade Federal do Rio Grande do Norte (UFRN).* Available in: https://lais.huol.ufrn.br/. Access date: May 29, 2022.

Lana RC, Paula AR, Silva AFS, et al. Validity of mHealth devices for counting steps in individuals with Parkinson's disease. *J Bodyw Mov Ther.* 2021;28:496–501. doi:10.1016/j.jbmt.2021.06.018.

Langhorne P, Bernhardt J, Kwakkel G. Stroke rehabilitation. *Lancet.* 2011;377(9778):1693–1702. doi:10.1016/S0140-6736(11)60325-5.

Lazzari RD, Politti F, Santos CA, et al. Effect of a single session of transcranial direct-current stimulation combined with virtual reality training on the balance of children with cerebral palsy: A randomized, controlled, double-blind trial. *Phys. Ther. Sci.* 2015;27(3):763–68. doi:10.1589/jpts.27.763.

Magrabi F, Ammenwerth E, McNair JB, et al. Artificial intelligence in clinical decision support: Challenges for evaluating ai and practical implications. *Yearb Med Inform.* 2019;28(1):128–34. doi:10.1055/s-0039-1677903.

MAISGOIÁS. Goiano que ficou tetraplégico em piscina faz tratamento inédito com aparelho suíço de R$ 2 milhões. 2022. Available in: https://www.maisgoias.com.br/goiano -que-ficou-tetraplegico-faz-tratamento-inedito-com-aparelho-suico-de-r-2-milhoes/. Access date: July 28, 2022.

Maniva SJCDF. Elaboração e validação de tecnologia educativa sobre acidente vascular cerebral para prevenção da recorrência. 2016. Available in: https://repositorio.ufc.br/handle /riufc/21580. Access date: July 28, 2022.

Martins SC, Weiss G, Almeida AG, et al. Validation of a smartphone application in the evaluation and treatment of acute stroke in a comprehensive stroke center. *Stroke.* 2020;51(1):240–246. doi:10.1161/STROKEAHA.119.026727.

Meirelles G. Inteligência artificial na saúde: Oportunidade ou ameaça? Available in: https:// mitsloanreview.com.br/post/inteligencia-artificial-na-saude-oportunidade-ou-ameaca. Access date: July 28, 2022.

Melo GEL, Kleiner AFR, Lopes JBP, et al. Effect of virtual reality training on walking distance and physical fitness in individuals with Parkinson's disease. *NeuroRehabilitation.* 2018;42(4):473–480. doi:10.3233/NRE-172355.

Mendes FAS, Pompeu JE, Lobo AM, et al. Motor learning, retention and transfer after virtual-reality-based training in Parkinson's disease–effect of motor and cognitive demands of games: A longitudinal, controlled clinical study. *Physiotherapy.* 2012;98(3):217–223. doi:10.1016/j.physio.2012.06.001.

Ministério da ciência, tecnologia e inovações (MCTI). *Estratégia brasileira de inteligência artificial-EBIA.* 2021. Available in: https://www.gov.br/mcti/pt-br/acompanhe-o-mcti /transformacaodigital/arquivosinteligenciaartificial/ebia-diagramacao_4-979_2021 .pdf. Access date: July 28, 2022.

Ministério da Saúde. *Secretaria de atenção primária à saúde. Linha de cuidado do acidente vascular cerebral (AVC) no adulto.* 2020. Available in: http://189.28.128.100/dab/docs/portaldab/publicacoes/LC_AVC_no_adulto.pdf. Access date: July 28, 2022.

Miranda CS, Oliveira TDP, Gouvêa JXM, et al. Balance training in virtual reality promotes performance improvement but not transfer to postural control in people with chronic stroke. *Games Health J.* 2019;8(4):294–300. doi:10.1089/g4h.2018.0075.

Moura JA, Chowdhury TI, Leal JC, et al. Virtual functional mobility test: A potential novel tool for assessing mobility of individuals with Parkinson's disease in a multitask condition. *J Clin Neurosci.* 2021;93:17–22. doi:10.1016/j.jocn.2021.08.017.

Muniz AMS, Liu H, Lyons KE, et al. Comparison among probabilistic neural network, support vector machine and logistic regression for evaluating the effect of subthalamic stimulation in Parkinson disease on ground reaction force during gait. *J Biomech.* 2010;43(4):720–726. doi:10.1016/j.jbiomech.2009.10.018.

Narváez F, Marín-Castrillón DM, Cuenca MC, et al. Development and implementation of technologies for physical telerehabilitation in Latin America: A systematic review of literature, programs and projects. *TecnoL.* 2017;20(40):155–176.

Nascimento LR, Fernandes MOP, Teixeira-Salmela LF, et al. Personal and organizational characteristics associated with evidence-based practice reported by Brazilian physical therapists providing service to people with stroke: A cross-sectional mail survey. *Braz J Phys Ther.* 2020;24(4):349–357. doi:10.1016/j.bjpt.2019.05.003.

NeuroRehab Directory. *Erigo.* Available in: https://www.neurorehabdirectory.com/rehab-products/erigo/. Access date: July 28, 2022a.

NeuroRehab Directory. *InMotion Arm.* Available in: https://www.neurorehabdirectory.com/rehab-products/inmotion/. Access date: July 28, 2022b.

Novo Hamburgo (NH). *Machine learning está em alta no Brasil, mas faltam profissionais.* 2022. Available in: https://www.jornalnh.com.br/informe_especial/2022/01/25/machine-learning-esta-em-alta-no-brasil-mas-faltam-profissionais.html. Access date: July 28, 2022.

Osborne JA, Botkin R, Colon-Semenza C, et al. Physical therapist management of parkinson disease: A clinical practice guideline from the American physical therapy association. *Phys Ther.* 2022;102(4):pzab302.

Ossanai DMT, Vitagliano E, Matuti GS, et al. utilização da robótica de membros superiores em pacientes pós acidente vascular cerebral crônico. In *Saberes e Competências em Fisioterapia 2.* Paraná: Atena Editora, 2019. doi:10.22533/at.ed.18719140421.

Pan American Health Organization (PAHO). *The burden of neurological conditions in the region of the Americas, 2000–2019.* 2021. Available in: https://www.paho.org/en/noncommunicable-diseases-and-mental-health/noncommunicable-diseases-and-mental-health-data-portal-3. Access date: July 28, 2022.

Paranashop. *Tecnologia. Óculos de realidade virtual tornam-se aliados na reabilitação de pacientes em hospital SUS de Curitiba.* 2022. Available in: https://paranashop.com.br/2022/03/oculos-de-realidade-virtual-tornam-se-aliados-na-reabilitacao-de-pacientes-em-hospital-sus-de-curitiba/. Access date: July 28, 2022.

Pedersen M, Verspoor K, Jenkinson M, et al. Artificial intelligence for clinical decision support in neurology. *Brain Commun.* 2020;2(2):fcaa096. doi:10.1093/braincomms/fcaa096.

Peretti A, Amenta F, Tayebati SK, et al. Telerehabilitation: Review of the state-of-the-art and areas of application. *JMIR Rehabil Assist Technol.* 2017;4(2):e7. doi:10.2196/rehab.7511.

Perez-Marcos D. Virtual reality experiences, embodiment, videogames and their dimensions in neurorehabilitation. *J Neuroeng Rehabil.* 2018;15(1):113. doi:10.1186/s12984-018-0461-0.

Pinhanez CS. C4AI: A new center for artificial intelligence to accelerate science and innovation in Brazil. 2020. Available in: https://www.ibm.com/blogs/research/2020/10/c4ai-a-new-center-for-artificial-intelligence-to-accelerate-science-and-innovation-in-brazil/. Acess date: July 28, 2022.

Polese JC, Faria GS, Ribeiro-Samora GA, et al. Google fit smartphone application or Gt3X Actigraph: Which is better for detecting the stepping activity of individuals with stroke? A validity study. *J Bodyw Mov Ther.* 2019;23(3):461–465. doi:10.1016/j.jbmt.2019.01.011.

Portal do Governo do Estado de São Paulo. *SP Notícias. Centro de reabilitação em São José dos Campos ganha laboratório de robótica.* 2014. Available in: https://setorsaude.com.br/santa-casa-adquire-equipamento-computadorizado-de-reabilitacao-robotica-inedito-na-america-latina/. Access date: July 28, 2022.

Portal Hospitais Brasil. Notícias. *Santa Casa de Porto Alegre adquire equipamento para reabilitação robótica de pacientes pós-Covid.* 2021. Available in: https://portalhospitaisbrasil.com.br/santa-casa-de-porto-alegre-adquire-equipamento-para-reabilitacao-robotica-de-pacientes-pos-covid/. Access date: July 28, 2022.

Queiroz O. *Como escolher o melhor óculos de realidade virtual do mercado?* 2022. Available in: https://www.terra.com.br/gameon/tech/como-escolher-o-melhor-oculos-de-realidade-virtual-do-mercado,e4605a58d4a77c9ca05eb7de6411e7ef23f6etv3.html. Access date: July 28, 2022.

RedeBrasilAVC. Available in: https://www.redebrasilavc.org.br/eventos-e-cursos/. Access date: July 28, 2022.

Santos MB, Oliveira CB, Santos A, et al. A comparative study of conventional physiotherapy versus robot-assisted gait training associated to physiotherapy in individuals with ataxia after stroke. *Behav Neurol.* 2018;2018:2892065. doi:10.1155/2018/2892065.

Santos WM, Siqueira AAG. Design and control of a transparent lower limb exoskeleton. *Biosyst Biorobotics.* 2018;175–179. doi:10.1007/978-3-030-01887-0_34.

Seat Mobile do Brasil (SMBrasil). Notícias. Novas tecnologias e robótica no processo de reabilitação. *Disease.* Available in: http://www.seatmobile.com.br/noticias/novas-tecnologias-e-robotica-no-processo-de-reabilitacao.html. Access date: July 28, 2022.

Setor Saúde. *Tecnologia e Inovação. Santa Casa adquire equipamento de reabilitação robótica inédito na América Latina.* 2018. Available in: https://setorsaude.com.br/santa-casa-adquire-equipamento-computadorizado-de-reabilitacao-robotica-inedito-na-america-latina/. Access date: July 28, 2022.

Silva TM, Costa LC, Costa LO. Evidence-based practice: A survey regarding behavior, knowledge, skills, resources, opinions and perceived barriers of Brazilian physical therapists from São Paulo state. *Braz J Phys Ther.* 2015;19(4):294–303. doi:10.1590/bjpt-rbf.2014.0102.

Silva-Couto MA, Siqueira AAG, Santos GL, et al. Ankle torque steadiness and gait speed after a single session of robot therapy in individuals with chronic hemiparesis: A pilot study. *Top Stroke Rehabil.* 2019;26(8):630–638. doi:10.1080/10749357.2019.1647984.

Teletime. *Brasil investe muito pouco em inteligência artificial na saúde.* 2020. Available in: https://teletime.com.br/08/12/2020/brasil-investe-muito-pouco-em-inteligencia-artificial-na-saude/. Access date: July 28, 2022.

Terranova TT, Simis M, Santos ACA, et al. Robot-Assisted therapy and constraint-induced movement therapy for motor recovery in stroke: Results from a randomized clinical trial. *Front Neurorobot.* 2021;15:684019. doi:10.3389/fnbot.2021.684019.

Universidade Federal do Rio Grande do Norte (UFRN). *Notícias. Com apoio do CNPq, UFRN desenvolve órtese para pacientes com ELA.* 2020. Available in: https://ufrn.br/imprensa/noticias/39423/com-apoio-do-cnpq-ufrn-desenvolve-ortese-para-pacientes

-com-ela#:~:text=Pesquisadores%20da%20Universidade%20Federal%20do,mais%20incapacitantes%20doen%C3%A7as%20neuromusculares%20conhecidas. Access date: July 28, 2022.

Vieira JLR, Ponte Junior LA. *An introduction to the internet of healthcare things.* Rio de Janeiro: Universidade Federal Fluminense, 2018. Available in: http://www.midiacom.uff.br/debora/images/disciplinas/2018-2/smm/trabalhos/IoHT-texto.pdf. Access date: July 28, 2022.

Winstein CJ, Stein J, Arena R, et al. Guidelines for adult stroke rehabilitation and recovery: A guideline for healthcare professionals from the American Heart Association/American Stroke Association. *Stroke.* 2016;47(6):e98–e169. doi:10.1161/STR.0000000000000098.

5 Virtual Reality and Magnetic Resonance Imaging of Anxious or Claustrophobic Patients

An Emerging Solution to a Longstanding Challenge?

D Hudson and C Heales

INTRODUCTION

Magnetic resonance imaging (MRI) has been one of the most significant developments within the field of medical diagnostics over the past 30 years, establishing itself as a key diagnostic tool within many aspects of modern-day medicine (Börnert & Norris, 2020; van Beek et al., 2019). Its benefits over other imaging techniques lie in its superior soft tissue contrast, multiplanar capability, and lack of ionizing radiation (Katti et al., 2019; Watson, 2015; Weidman et al., 2015). Within the UK, demand for the modality continues to follow an upward trend (NHS England, 2020; The Royal College of Radiologists, 2017) and is set to increase further as a result of delays and increased waiting lists arising following the COVID-19 pandemic and the lockdowns of 2020 (Richards, 2020).

As an imaging technique, MRI involves the use of a strong magnetic field with exposure to radiofrequency waves in order to generate an image (Katti et al., 2019). In the vast majority of cases, the MRI scanner is composed of a superconducting magnet, which, by its nature, is cylindrical in shape. Patients are required to lay still within its bore for the duration of their scans with receiver coils placed over the part of the body being scanned in order to obtain an image (Brunnquell et al., 2020).

MRI AND CLAUSTROPHOBIA

Since its early days in clinical practice, magnet technology and overall scanner design have evolved and scanners now appear more open; the bore is typically much shorter in length and bore diameters are now up to 70 cm (Brunnquell et al., 2020).

DOI: 10.1201/9781003272786-5

Advances in scanning software (sequences) and coil technology have also helped reduce scan acquisition times so that the time spent within the scanner is often less than before (Börnert & Norris, 2020; Brunnquell et al., 2020). In other words, scanners are becoming more patient friendly by design, coupled with technological developments that reduce overall scan durations (Ahlander et al., 2020; Brunnquell et al., 2020; Dewey et al., 2007; Iwan et al., 2020).

As a consequence, it has been argued that these more patient-friendly designs, together with the accompanying technological enhancements, would reduce or even remove the need to consider claustrophobia-related anxiety in MRI as they become commonplace (Wood & McGlynn, 2000). However, this has not come to pass. Firstly, MRI scanners have a reasonable life span; for example, as of 2017, over 50% of scanners in the UK were still traditional narrow-bore (60 cm) systems (The Royal College of Radiologists, 2017), with scanners over 10 years old still being used in practice (European Coordination Committee of the Radiological Electromedical and Healthcare IT industry, 2021). Furthermore, technological advances enable existing magnet technology to be upgraded. This is cost-effective for clinical departments but can mean that the traditional narrow-bore scanner is maintained (GE Healthcare, 2019). Secondly, current service evaluation data demonstrate that sufficient numbers of patients still experience anxiety around MRI (Hudson et al., 2022). Hence, 20 years on, the issue of claustrophobia, and therefore the impact on MRI scan outcomes, is still very much relevant in clinical practice.

The fear or anxiety most commonly presenting in MRI scanners is claustrophobia. This is categorized as a situational phobia within the Diagnostic and Statistical Manual of mental disorders (American Psychiatric Association, 2015), being a phobia of enclosed spaces, in this instance associated with the confines of the MRI scanner. A further consideration is cleithrophobia – the fear of being trapped and unable to escape (Panic and Anxiety Community Suppport, 2020) – which, over time, can become synonymous with claustrophobia for some people (Radomsky et al., 2001). As with any phobia, the outcome is an irrational, disproportionate response which affects an individual's ability to cope in that particular situation (American Psychiatric Association, 2015; Buchanan & Coulson, 2012).

Fear of enclosed spaces has been reported in 2.2% of the general population (Wardenaar et al., 2017), with the reported incidence rates in the literature of scan-related claustrophobia varying (Munn et al., 2014). MRI services will typically measure claustrophobia in terms of the impact upon scan completion, with a recent study placing the rate of incomplete examinations at less than 1% (Hudson et al., 2022). However, MRI radiographers themselves report almost daily occurrences of interacting with patients who are visibly or communicably anxious or claustrophobic (Al-Shemmari et al., 2022; Hudson et al., 2022), with little evidence of the rate of incidence changing over the past decades (Tischler et al., 2008). It is also important to recognize that claustrophobia may be more common than is realized if the typical measures used by clinical services are observable effects and/or incomplete or abandoned scans. It is known that claustrophobic responses range from feeling anxious through to extreme panic (Munn & Jordan, 2011). In other words, it is not necessarily obvious to what degree someone may be experiencing a sense of claustrophobia, particularly if the impacted individual manages to control their response and cope, showing minimal to no signs of distress – people may seemingly cope while still struggling with the experience.

THE IMPACT OF EXPERIENCING CLAUSTROPHOBIA

The implications for individuals of experiencing claustrophobia or extreme anxiety when undergoing an MRI scan are variable; for some, it will mean not being able to tolerate any scan at all (Bangard et al., 2007; Eshed et al., 2007; Munn et al., 2015; Napp et al., 2017; Norbash et al., 2016), while, for others, it may result in the examination being abandoned part way through, or for those able to tolerate being within the scanner there may still be challenges around patient movement and degradation of image quality (Bangard et al., 2007; Klaming et al., 2015; Nguyen et al., 2020; Powell et al., 2015; Sadigh et al., 2017). Finally, some patients may complete their scan successfully but really dislike the experience, which can then impact on any future scan attendance or their wider interaction with other health services (Lloyd, 2020).

Ultimately, a degraded or incomplete MRI scan, or an inability to undergo MR imaging at all, can all potentially have an adverse impact on the patient's diagnosis and onward management (Munn et al., 2014). For example, although there are sometimes alternative imaging techniques that can be used instead, organizing these can create delays, and the alternative may not be able to provide as much information as MRI. In addition, for any patient who struggles in this way, there is the possibility of them experiencing heightened anxiety should they need any further MRI scans in the future (Enders et al., 2011). This clearly has a negative impact upon their overall experience, and potentially on their general well-being. From a service delivery perspective, as already mentioned, there can be adverse impacts on health outcomes, and there may be cost and administrative implications for organizing repeat attempts at MRI and/or arranging alternative imaging or other tests. There can also be cost and productivity implications for the MRI scanner service itself, leading to lost revenue, lost scanner time, and reduced throughput (Andre et al., 2015; Dewey et al., 2007; Enders et al., 2011; Nguyen et al., 2020; Norbash et al., 2016), which is of particular concern in the current climate, for example, in the UK where there are significant numbers of people already waiting for imaging combined with a lack of scanner capacity (Richards, 2020; The Royal College of Radiologists, 2017).

USING VIRTUAL REALITY TO SUPPORT PATIENTS REFERRED FOR MRI

Virtual reality (VR) is defined as "a computer-generated digital environment that can be experienced and interacted with as if that environment were real" (Jerald, 2016, p. 9). This is most commonly achieved by a user wearing a head-mounted device with three-dimensional (3D) displays that transport the wearer's awareness away from the physical world and into the 3D simulation. It is this truly immersive experience that will be referred to as VR in this chapter. Although predominantly developed for use in the field of gaming, VR technology is increasingly being used for other applications. For example, it can be used to provide a safe space for high-risk training, with examples being use by the military or to help prepare journalists for working within hostile environments (McIntosh, 2022).

There are two essential components needed for VR to be effective, those of presence and immersion. Presence relates to how much the user feels they are really within the simulated environment, whereas immersion relates to feelings of engagement (Jerald, 2016; Spiegel, 2020; Wiederhold & Bouchard, 2014). An ability to interact with the environment can also be relevant in particular applications, such as training VR environments, and also increases the sense of immersion and presence (Mütterlein, 2018). VR is also increasingly being used in a therapeutic setting, such as for pain management and psychological interventions (Spiegel, 2020). The growth in published literature from the early 1990s (Pawassar & Tiberius, 2021) onwards indicates five principle uses of VR, namely; as exposure therapy, distraction during procedures, motivation during rehabilitation, measurement of ability, and patient engagement (Rizzo & Koenig, 2017). Specifically in relation to MRI, the use of VR is increasingly being explored in a number of ways and contexts, including exposure therapy, patient engagement, as a distraction technique, and for staff training and development.

VR AS EXPOSURE THERAPY

Traditional approaches to treating phobias are through progressive exposure to the stimulus that triggers the symptoms, either in real-life or through visualization (Mazurek et al., 2019), until the distressed response decreases (Freitas et al., 2021). VR is able to offer controllable progression in intensity (in terms of exposure) from within a safe space (Malbos et al., 2008; Wiederhold & Bouchard, 2014). The VR environment enables people to slowly develop tolerance to the feelings of discomfort and anxiety associated with the trigger in a completely controlled way, with the ability to stop at any point, and is being shown to be effective for the treatment of most phobias (Freitas et al., 2021). However, VR never really gained momentum as a mainstream tool for supporting people through their MRI scans until more recently. In part, this was because, in the late 1990s and early 2000s, it was not feasible for bespoke MRI environments to be created and used for VR simulation. Hence, early use of VR involved the concepts of rooms or lifts that decreased in size. However, this did enable the principle of exposure therapy (repeated exposures over a period of time) to be explored, and ultimately this was seen to be effective in terms of enabling people to undergo treatment for both computed tomography (CT) and MRI (Botella et al., 1998; Wiederhold & Wiederhold, 2000). Now, of course, with the significant advances in processing power and increasing affordability, it is possible to model 3D virtual replicas of the MRI scan room and environment and, using headsets, to immerse people fully in this recreated clinical environment, even to the point of letting them explore in their own way and in their own time (Brown et al., 2018; Nakarada-Kordic et al., 2020). Comparisons between the use of these more recent iterations of MRI-VR environments (VRE) and "mock" scanners have shown that VREs can now adequately replicate the MRI experience (Nakarada-Kordic et al., 2020). Hence, the evolving capabilities of the technology have reignited interest in the use of VREs, combined with approaches such as exposure therapy, to help prepare patients with claustrophobia for their MRI scans.

VR FOR PATIENT ENGAGEMENT

Another potentially straightforward use of VR is in terms of simply enabling people to engage with, and therefore familiarize themselves with, the MRI environment. Aspects as simple as rehearsing entry into and exit from an actual, physical scanner (Enders et al., 2011; Funk et al., 2014) and lying within the scanner (Munn & Jordan, 2011) have all been shown to be beneficial for people with claustrophobia, but it can be difficult for clinical services to provide sufficient time on clinical scanners due to the pressures of day-to-day service provision (Hudson et al., 2022). VR can provide an opportunity for this type of familiarization without taking time away from the clinical service. Another benefit includes the potential for VR to be made available to patients within their own homes (Nakarada-Kordic et al., 2020).

VR AS A DISTRACTION TECHNIQUE

VR environments that don't replicate the MRI scanner and environment can also be used as straightforward distraction techniques. As with using VR for familiarization, the use of these wouldn't provide a therapeutic benefit, but, provided the equipment is MR conditionally safe (that is, is able to be used within the MRI scanner), this may provide a more complete distraction during the scan itself than music alone (Garcia-Palacios et al., 2007). Certainly, for children, playing games during their scan may reduce stress and anxiety and avoid the need for a medicated approach (sedation or general anesthesia) (Liszio et al., 2020) while also being of benefit to adults (Qian et al., 2021).

VR FOR STAFF DEVELOPMENT

VR can also, potentially, be used for staff training to help them better understand the patient experience (Hudson & Heales, 2023). Using VR, so that staff can experience a healthcare episode from the perspective of the patient, has been shown to elicit empathy (Brydon et al., 2021), which should result in improved patient outcomes in real-world clinical practice (Sanders et al., 2021), although more research is needed. VR can also be used to present scenarios in which staff can rehearse their interactions with patients (Brydon et al., 2021). This type of VR-based patient interaction training has potential benefits over traditional role play due to the absence of external observers (Sapkaroski et al., 2022), who may inhibit the trainee. Of course, there are challenges in creating such content in order to achieve realism and to introduce suitable interactivity, such as to replicate the experience of interacting with another person. Nevertheless, it is clear that having confidence in their own ability to manage and support anxious patients is an important skill for radiographers (Al-Shemmari et al., 2022), and being able to use VR as a training tool could really accelerate the development of radiographer skills ready for practice.

It can be seen, therefore, that there is a range of ways in which VR could play a role in enhancing the patient experience in MRI. However, these approaches are not mutually exclusive and various elements can be combined, for example, to support children through an MRI without using sedation or general anesthesia. "Mock" scanners specifically designed for children, i.e., play simulators, are available and have been shown to be effective in terms of both scan outcome and cost, when compared with using general anesthesia for scanning young children (Heales & Lloyd, 2022). However, such mock scanners are expensive to procure, require space, and typically are not widely available (Chapman et al., 2010; Thorpe et al., 2008). VR has been shown to be as effective as the use of a mock scanner (Stunden et al., 2021) and so the use of this technology may provide a viable alternative for more departments, with approaches ranging from full immersion in the VRE using 360° video (Ashmore et al., 2019; Stunden et al., 2021) through to combining exposure therapy with the use of preparatory information, gamification, and play therapy (Liszio et al., 2020; Liszio & Masuch, 2017). Gamification can include the use of storytelling and the incorporation of puzzle or competition elements, and rewards and can be tailored to the age of the child (Liszio et al., 2020; Liszio & Masuch, 2017). There is also scope for the parents/guardians to be able to undergo the same preparatory VR experiences as their child so as to be able to better understand and support their child through the MRI scan procedure (Stunden et al., 2021). Benefits of such an approach would also include reduced need to use sedation or general anesthesia with the associated clinical risks and service delivery costs.

The use of VR to negate the need to use medication may also be relevant to adults. Although radiographers tend to regard the use of drugs such as anxiolytics to be an effective approach to support adult patients through MRI (Al-Shemmari et al., 2022; Hudson et al., 2022; Tischler et al., 2008), the use of such drugs in this context is being discouraged due to concerns around safety (The Royal College of Radiologists, 2018); using VR in any of the ways described here to reduce the use of pharmacological approaches for any age of patient is clearly of benefit.

POSSIBLE PSYCHOLOGICAL AND/OR THERAPEUTIC MECHANISMS OF ACTION OF VR

The use of VR can help people undergo imaging, particularly MRI. It would also appear that there are high levels of acceptance of the use of VR among the patient groups themselves, which is key if the use of this technology is to become more widespread. The current question still needing to be explored, though, is around possible mechanisms of action; specifically, is there some psychological and/or therapeutic change occurring or is it as simple as being able to familiarise with the environment and take time to prepare with a supportive healthcare practitioner?

It is likely that a number of different mechanisms underpin what radiographers/ MRI technologists perceive to be claustrophobia. Anxiety is defined as "an emotion characterized by feelings of tension, worried thoughts and physical changes like increased blood pressure" (American Psychiatric Association, n.d.), whereas fear is generally felt in response to a present danger and so may manifest when a patient first sees the MRI scanner, and anxiety is a natural response associated with what an individual perceives as an upcoming threat or source of stress (Booth & Bell, 2013).

How patients respond to anxiety and/or fear will vary. The expectancy model of fear (Reiss, 1991) suggests that how people respond to situations they fear depends on their beliefs about what the most likely outcome will be. Furthermore, how someone responds to fear will also vary depending on their personal ability to cope with the particular situation (Bandura, 1988). In other words, if someone feels anxiety and/or fear but believes they are able to cope, then they perceive the situation as a challenge; conversely, if they don't feel able to cope, they perceive the situation as a threat (Blascovich, 2008). If perceiving the situation as a threat, people then tend to fall into one of two behaviors, namely worry or avoidance (American Psychiatric Association, 2015; Brown, 2021). Avoidance, in the context of MRI, could mean a patient not attending, not starting, or not completing their scan. However, if able to perceive the MRI scan as a challenge, then an individual is better motivated to face their fear, resulting in a successful imaging outcome. This is known as the biopsychosocial model of challenge and threat (Blascovich, 2008; Tomaka et al., 1993) and it is this mechanism that is thought likely to underpin the use of VR in relation to the management of phobias through exposure therapy and acclimatization. VR is used as an alternative to *in vivo* exposure and as an adjunct to the delivery of cognitive behavioral therapy (Wiederhold & Bouchard, 2014). The use of VREs allows patients to experience fear-inducing stimuli while supported in a safe space. The key element is repeated exposure aiming to induce a response and aid therapy over repeated sessions (Botella et al., 2000) allowing successful completion of a scan.

Early studies within the context of medical imaging were based on this type of exposure therapy, although researchers weren't able to use a realistic imaging environment. As previously described, early attempts utilized "shrinking" lift or room scenarios. Nevertheless, by using psychologically validated questionnaires and scales, it was shown that the exposure therapy (with or without supplementary techniques, such as breathing exercises) did reduce people's discomfort in relation to being in increasingly smaller spaces (Wiederhold & Wiederhold, 2000). This approach is now being revisited, initially using the lift scenario supplemented by video footage of the MRI scanner (Rahani et al., 2018), but increasingly through modeled VREs of the scanner room and environment itself (Brown et al., 2018; Nakarada-Kordic et al., 2020). The usefulness of VR in this very particular context is becoming established, although there is scope for further research to understand any therapeutic mechanisms and how best to implement them.

FUTURE DIRECTIONS AND LIMITATIONS OF VR AS A SUPPORT TOOL IN MRI

It is important to be able to understand whether VR has a therapeutic element and how it can be used to best effect, in relation to supporting patients through MRI as there are cost, space, training, and time implications for adding a VRE component to a clinical service. A further, currently unanswered question is whether "any" realistic MRI scan environment is sufficient, or does the VRE need to resemble the "actual" scanner that the patient will have their scan on as it is known that the fidelity of the VRE, compared with the real-world equivalent, is thought to be important (Rizzo & Koenig, 2017).

Likewise, if, as early research suggests, there are alterations in recorded anxiety/claustrophobia scales as a consequence of time spent in the VRE, does this mean that VRE tools can be used to help patients overcome their anxiety/claustrophobia on a more permanent basis? This is an area of emerging research, and the outcomes of such research will help MRI departments determine whether to invest in the hardware, software, and staff development needed to effectively incorporate VR within their service provision without taking up time on busy clinical scanners.

Currently support strategies offered by MRI departments tend to be targeted at those patients who struggle to complete their MRI scans and is unlikely to take into account potentially up to 25% of patients who are claustrophobic and/or anxious and yet cope with their scans (Busacchio et al., 2021; Oliveri et al., 2018). Hence, using scan incompletion due to claustrophobia as a criterion for intervention is unlikely to be an adequate measure of high-quality patient care (Hyde & Hardy, 2021). There is scope for a discussion about making technologies such as VR available to any patient who feels they would benefit, even if they do tolerate the scans.

There is also scope for further exploration of the types of anxiety associated with MRI. There are suggestions within the literature that people with particular health conditions (typically severe and/or chronic) may display above-average levels of concern about MRI. Although this can be in relation to the scan itself (Katznelson et al., 2008), it is suggested that anxiety, in some at least, is related to receiving the results (Engels et al., 2019). It is therefore important that MRI radiographers/technologists don't assume all anxiety is related to the scan procedure itself. The use of questionnaires, such as the claustrophobic questionnaire or anxiety sensitivity index, have shown promise as useful predictive tools (Booth & Bell, 2013; Mohlman et al., 2012; Napp et al., 2017) that would enable VR resources to be targeted appropriately (Hudson et al., 2022). Perhaps moving forward, this is where the use of artificial intelligence could play a role, drawing on individual data and using algorithms to enhance patient-centered services through providing individualized scheduling and tailored information.

It is also important to come back to the interpersonal interactions that underpin all healthcare provisions. One common feature of all data on anxiety and claustrophobia incidence rates is high variability between different MRI sites (Hudson et al., 2022), suggesting that other factors are at play (Carlsson & Carlsson, 2013). Although this could be department design/atmosphere, a likely explanation is the need for the radiographer/MRI technologist to have sufficient time to support each patient (Hudson et al., 2022). Targeted training has also been shown to reduce rates of abandoned scans (Lang et al., 2010; Norbash et al., 2016). In other words, technology, such as VR, should be seen as a supplement to the provision of patient-centered care by the healthcare personnel working within MRI services.

In conclusion, it can be seen that there is real potential for VR to be used as a tool to support individuals going through MRI in a number of ways, including preparation for a scan, distraction during a scan, and potentially as a therapeutic tool to increase the ability to cope with claustrophobia/anxiety. VR packages can also be designed for target populations, such as children, as well as for adults. The emerging evidence base discussed here certainly indicates that there is real-world potential for this technology, with its ability to reasonably replicate the MRI environment having a positive impact on patient experience. Although not really explored in relation to MRI, there is also the potential for the use of VR as a teaching/staff development tool. Hence, further research into the use of this technology in each and all of these different ways is certainly warranted. Finally, as VR technology is highly likely to continue to evolve and improve, its applications as a patient support and training tool will likely further develop too.

REFERENCES

Ahlander, B. M., Engvall, J., & Ericsson, E. (2020). Anxiety during magnetic resonance imaging of the spine in relation to scanner design and size. *Radiography*, *26*(2), 110–116. https://doi.org/10.1016/j.radi.2019.09.003

Al-Shemmari, A. F., Herbland, A., Akudjedu, T. N., & Lawal, O. (2022). Radiographer's confidence in managing patients with claustrophobia during magnetic resonance imaging. *Radiography*, *28*(1), 148–153. https://doi.org/10.1016/j.radi.2021.09.007

American Psychiatric Association. (n.d.). *Anxiety.* Retrieved August 31, 2022, from https://www.apa.org/topics/anxiety

American Psychiatric Association. (2015). *Anxiety Disorders: DSM-5® Selections.* https://doi.org/10.1007/978-1-4899-1498-9_5

Andre, J. B., Bresnahan, B. W., Mossa-Basha, M., Hoff, M. N., Patrick Smith, C., Anzai, Y., & Cohen, W. A. (2015). Toward quantifying the prevalence, severity, and cost associated with patient motion during clinical MR examinations. *Journal of the American College of Radiology*, *12*(7), 689–695. https://doi.org/10.1016/j.jacr.2015.03.007

Ashmore, J., Di Pietro, J., Williams, K., Stokes, E., Symons, A., Smith, M., … McGrath, C. (2019). A free virtual reality experience to prepare pediatric patients for magnetic resonance imaging: Cross-sectional questionnaire study. *Journal of Medical Internet Research: Pediatrics and Parenting*, *2*(1), e11684; 1–10. https://doi.org/10.2196/11684

Bandura, A. (1988). Self-efficacy conception of anxiety. *Anxiety Research*, *1*(2), 77–98. https://doi.org/10.1080/10615808808248222

Bangard, C., Paszek, J., Berg, F., Eyl, G., Kessler, J., Lackner, K., & Gossmann, A. (2007). MR imaging of claustrophobic patients in an open 1.0 T scanner: Motion artifacts and patient acceptability compared with closed bore magnets. *European Journal of Radiology*, *64*(1), 152–157. https://doi.org/10.1016/j.ejrad.2007.02.012

Blascovich, J. (2008). Challenge and threat. In A. J. Elliot (Ed.), *Handbook of Approach and Avoidance Motivation* (pp. 431–444). New York: Psychology Press - Taylor and Francis Group.

Booth, L., & Bell, L. (2013). Screening for claustrophobia in MRI – a pilot study. *European Scientific Journal*, *9*(18), 1857–7881.

Börnert, P., & Norris, D. G. (2020). A half-century of innovation in technology-preparing MRI for the 21st century. *The British Journal of Radiology*, *93*(1111), 20200113. https://doi.org/10.1259/bjr.20200113

Botella, C, Bafios, R. M., Perpina, C., Villa, H., Alcafiiz, M., & Rey, A. (1998). Short Communication Virtual reality treatment of claustrophobia: a case report. *Behaviour Research and Therapy*, *36*, 239–246.

Botella, C., Baños, R. M., Villa, H., Perpiñá, C., & García-Palacios, A. (2000). Virtual reality in the treatment of claustrophobic fear: A controlled, multiple-baseline design. *Behavior Therapy*, *31*(3), 583–595. https://doi.org/10.1016/S0005-7894(00)80032-5

Brown, B. (2021). *Atlas of the Heart*. London: Penguin Rsndom House.

Brown, R. K. J., Petty, S., O'Malley, S., Stojanovska, J., Davenport, M. S., Kazerooni, E. A., & Fessahazion, D. (2018). Virtual reality tool simulates MRI experience. *Tomography*, *4*(3), 95–98. https://doi.org/10.18383/j.tom.2018.00023

Brunnquell, C. L., Hoff, M. N., Balu, N., Nguyen, X. V., Oztek, M. A., & Haynor, D. R. (2020). Making magnets more attractive: Physics and engineering contributions to patient comfort in MRI. *Topics in Magnetic Resonance Imaging : TMRI*, *29*(4), 167–174. https://doi.org/10.1097/RMR.0000000000000246

Brydon, M., Kimber, J., Sponagle, M., MacLaine, J., Avery, J., Pyke, L., & Gilbert, R. (2021). Virtual reality as a tool for eliciting empathetic behaviour in carers: An integrative review. *Journal of Medical Imaging and Radiation Sciences*, *52*(3), 466–477. https://doi.org/10.1016/j.jmir.2021.04.005

Buchanan, H., & Coulson, N. (2012). *Phobias*. Basinsgtoke: Palgrave Macmillan.

Busacchio, D., Mazzocco, K., Gandini, S., Pricolo, P., Masiero, M., Summers, P. E., ... Petralia, G. (2021). Preliminary observations regarding the expectations, acceptability and satisfaction of whole-body MRI in self-referring asymptomatic subjects. *The British Journal of Radiology*, *94*(1118), 20191031. https://doi.org/10.1259/bjr.20191031

Carlsson, S., & Carlsson, E. (2013). "The situation and the uncertainty about the coming result scared me but interaction with the radiographers helped me through": A qualitative study on patients' experiences of magnetic resonance imaging examinations. *Journal of Clinical Nursing*, *22*(21–22), 3225–3234. https://doi.org/10.1111/jocn.12416

Chapman, H. A., Bernier, D., & Rusak, B. (2010). MRI-related anxiety levels change within and between repeated scanning sessions. *Psychiatry Research - Neuroimaging*, *182*(2), 160–164. https://doi.org/10.1016/j.pscychresns.2010.01.005

Dewey, M., Schink, T., & Dewey, C. F. (2007). Claustrophobia during magnetic resonance imaging: Cohort study in over 55,000 patients. *Journal of Magnetic Resonance Imaging*, *26*(5), 1322–1327. https://doi.org/10.1002/jmri.21147

Enders, J., Zimmermann, E., Rief, M., Martus, P., Klingebiel, R., Asbach, P., ... Dewey, M. (2011). Reduction of claustrophobia with short-bore versus open magnetic resonance imaging: A randomized controlled trial. *PLoS ONE*, *6*(8), 1–10. https://doi.org/10.1371/journal.pone.0023494

Engels, K., Schiffmann, I., Weierstall, R., Rahn, A. C., Daubmann, A., Pust, G., ... Heesen, C. (2019). Emotions towards magnetic resonance imaging in people with multiple sclerosis. *Acta Neurologica Scandinavica*, *139*(6), 497–504. https://doi.org/10.1111/ane.13082

Eshed, I., Althoff, C. E., Hamm, B., & Hermann, K. G. A. (2007). Claustrophobia and premature termination of magnetic resonance imaging examinations. *Journal of Magnetic Resonance Imaging*, *26*(2), 401–404. https://doi.org/10.1002/jmri.21012

European Coordination Committee of the Radiological Electromedical and Healthcare IT Industry. (2021). *Medical Imaging Equipment Age Profile and Denisty 2021 Edition*. https://www.cocir.org/fileadmin/Publications_2021/COCIR_Medical_Imaging_Equipment_Age_Profile_Density_-_2021_Edition.pdf

Freitas, J. R. S., Velosa, V. H. S., Abreu, L. T. N., Jardim, R. L., Santos, J. A. V., Peres, B., & Campos, P. F. (2021). Virtual reality exposure treatment in phobias: A systematic review. *Psychiatric Quarterly*, *92*(4), 1685–1710. https://doi.org/10.1007/s11126-021-09935-6

Funk, E., Thunberg, P., & Anderzen-Carlsson, A. (2014). Patients' experiences in magnetic resonance imaging (MRI) and their experiences of breath holding techniques. *Journal of Advanced Nursing, 70*(8), 1880–1890. https://doi.org/10.1111/jan.12351

Garcia-Palacios, A., Hoffman, H. G., Richards, T. R., Seibel, E. J., & Sharar, S. R. (2007). Use of virtual reality distraction to reduce claustrophobia symptoms during a mock magnetic resonance imaging brain scan: A case report. *Cyberpsychology and Behavior, 10*(3), 485–488. https://doi.org/10.1089/cpb.2006.9926

GE Healthcare. (2019). *The Guide to Upgrading an MRI Scanner.* https://www.gehealthcare.com/-/jssmedia/gehc/us/files/products/magnetic-resonance-imaging/mri-upgrades-and-lifecycle-guide.pdf?rev=-1

Heales, C. J., & Lloyd, E. (2022). Play simulation for children in magnetic resonance imaging. *Journal of Medical Imaging and Radiation Sciences, 53*(1), 10–16. https://doi.org/10.1016/j.jmir.2021.10.003

Hudson, D. M., & Heales, C. (2023), "I think this could be a big success"–A mixed methods study on practitioner perspectives on the acceptance of a virtual reality tool for preparation in MRI. *Radiography, 29*(5), 851–861.

Hudson, D. M., Heales, C., & Meertens, R. (2022). Review of claustrophobia incidence in MRI: A service evaluation of current rates across a multi-centre service. *Radiography.* https://doi.org/10.1016/j.radi.2022.02.010

Hudson, D. M., Heales, C., & Vine, S. J. (2022). Radiographer perspectives on current occurrence and management of claustrophobia in MRI. *Radiography, 28*(1), 154–161. https://doi.org/10.1016/j.radi.2021.09.008

Hyde, E., & Hardy, M. (2021). Patient centred care in diagnostic radiography (Part 1): Perceptions of service users and service deliverers. *Radiography, 27*(1), 8–13. https://doi.org/10.1016/j.radi.2020.04.015

Iwan, E., Yang, J., Enders, J., Napp, A. E., Rief, M., & Dewey, M. (2020). Patient preferences for development in MRI scanner design: A survey of claustrophobic patients in a randomized study. *European Radiology.* https://doi.org/10.1007/s00330-020-07060-9

Jerald, J. (2016). *The VR Book: Human-Centered Design for Virtual Reality* (First). Association for Computing Machinery and Morgan & Claypool. https://doi.org/10.1145/2792790

Katti, G., Ara, S. A., & Shireen, A. (2019). Magnetic resonance imaging (MRI) - A review. *International Journal of Dental Clinics, 3*(1), 65–70.

Katznelson, R., Djaiani, G. N., Minkovich, L., Fedorko, L., Carroll, J., Borger, M. A., … Karski, J. (2008). Prevalence of claustrophobia and magnetic resonance imaging after coronary artery bypass graft surgery. *Neuropsychiatric Disease and Treatment, 4*(2), 487–493. https://doi.org/10.2147/ndt.s2699

Klaming, L., Van Minde, D., Weda, H., Nielsen, T., & Duijm, L. E. M. (2015). The relation between anticipatory anxiety and movement during an MR examination. *Academic Radiology, 22*(12), 1571–1578. https://doi.org/10.1016/j.acra.2015.08.020

Lang, E. V., Ward, C., & Laser, E. (2010). Effect of team training on patients' ability to complete MRI examinations. *Academic Radiology, 17*(1), 18–23. https://doi.org/10.1016/j.acra.2009.07.002

Liszio, S., Basu, O., & Masuch, M. (2020). A universe inside the MRI scanner: An in-bore virtual reality game for children to reduce anxiety and stress. In *CHI PLAY'20* (pp. 46–57). https://doi.org/10.1145/3410404.3414263

Liszio, S., Graf, L., Basu, O., & Masuch, M. (2020). Pengunaut trainer: A playful VR app to prepare children for MRI examinations: In-depth game design analysis. *Proceedings of the Interaction Design and Children Conference, IDC 2020*, 470–482. https://doi.org/10.1145/3392063.3394432

Liszio, S., & Masuch, M. (2017). Virtual reality MRI: Playful reduction of children's anxiety in MRI exams. In *IDC 2017 - Proceedings of the 2017 ACM Conference on Interaction Design and Children* (pp. 127–136). https://doi.org/10.1145/3078072.3079713

Lloyd, L. (2020). A tale of two MRIs. *Journal of Medical Imaging and Radiation Sciences*, *51*(4), S9–S10. https://doi.org/10.1016/j.jmir.2020.05.010

Malbos, E., Mestre, D. R., Note, I. D., & Gellato, C. (2008). Virtual reality and claustrophobia: Multiple components therapy involving game editor virtual environments exposure. *Cyberpsychology and Behavior*, *11*(6), 695–697. https://doi.org/10.1089/cpb.2007.0246

Mazurek, J., Kiper, P., Cieślik, B., Rutkowski, S., Mehlich, K., Turolla, A., & Szczepańska-Gieracha, J. (2019). Virtual reality in medicine: A brief overview and future research directions. *Human Movement*, *20*(3), 16–22. https://doi.org/10.5114/hm.2019.83529

McIntosh, V. (2022). Dialing up the danger: Virtual reality for the simulation of risk. *Frontiers in Virtual Reality, August*, 1–17. https://doi.org/10.3389/frvir.2022.909984

Mohlman, J., Eldreth, D. A., Price, R. B., Chazin, D., & Glover, D. A. (2012). Predictors of unsuccessful magnetic resonance imaging scanning in older generalized anxiety disorder patients and controls. *Journal of Behavioral Medicine*, *35*(1), 19–26. https://doi.org/10.1007/s10865-011-9326-8

Munn, Z., & Jordan, Z. (2011). The patient experience of high technology medical imaging: A systematic review of the qualitative evidence. *Radiography*, *17*(4), 323–331. https://doi.org/10.1016/j.radi.2011.06.004

Munn, Z., Moola, S., Lisy, K., Riitano, D., & Murphy, F. (2014). Claustrophobia in magnetic resonance imaging: A systematic review and meta-analysis. *Radiography*, *21*(2), e59–e63. https://doi.org/10.1016/j.radi.2014.12.004

Munn, Z., Pearson, A., Jordan, Z., Murphy, F., Pilkington, D., & Anderson, A. (2015). Patient anxiety and satisfaction in a magnetic resonance imaging department: Initial results from an action research study. *Journal of Medical Imaging and Radiation Sciences*, *46*(1), 23–29. https://doi.org/10.1016/j.jmir.2014.07.006

Mütterlein, J. (2018). The three pillars of virtual reality? Investigating the roles of immersion, presence, and interactivity. *Proceedings of the Annual Hawaii International Conference on System Sciences*, 2018-Janua, 1407–1415. https://doi.org/10.24251/hicss.2018.174

Nakarada-Kordic, I., Reay, S., Bennett, G., Kruse, J., Lydon, A. M., & Sim, J. (2020). Can virtual reality simulation prepare patients for an MRI experience? *Radiography*, *26*(3), 205–213. https://doi.org/10.1016/j.radi.2019.11.004

Napp, A. E., Enders, J., Roehle, R., Diederichs, G., Rief, M., Zimmermann, E., … Dewey, M. (2017). Analysis and prediction of claustrophobia during MR imaging with the claustrophobia questionnaire: An observational prospective 18-month single-center study of 6500 patients. *Radiology*, *283*(1), 148–157. https://doi.org/10.1148/radiol.2016160476

Nguyen, X. V., Tahir, S., Bresnahan, B. W., Andre, J. B., Lang, E. V., Mossa-Basha, M., … Bourekas, E. C. (2020). Prevalence and financial impact of claustrophobia, anxiety, patient motion, and other patient events in magnetic resonance imaging. *Topics in Magnetic Resonance Imaging: TMRI*, *29*(3), 125–130. https://doi.org/10.1097/RMR.0000000000000243

NHS England. (2020). *Diagnostic Imaging Dataset Annual Statistical Statistical Release 2019/20*. NHS England. Retrieved from http://www.wjgnet.com/1949-8470/full/v6/i1/1.htm

Norbash, A., Yucel, K., Yuh, W., Doros, G., Ajam, A., Lang, E., … Mayr, N. (2016). Effect of team training on improving MRI study completion rates and no-show rates. *Journal of Magnetic Resonance Imaging*, *44*(4), 1040–1047. https://doi.org/10.1002/jmri.25219

Oliveri, S., Paola, P., Silvia, P., Faccio, F., Valentina, L., Paul, S., … Gabriella, P. (2018). Investigating cancer patient acceptance of whole body MRI. *Clinical Imaging*, *52*(July), 246–251. https://doi.org/10.1016/j.clinimag.2018.08.004

Panic and Anxiety Community Suppport. (2020). *Claustrophobia-and-cleithrophobia*. Retrieved from https://panicandanxiety.org/claustrophobia-and-cleithrophobia/

Pawassar, C. M., & Tiberius, V. (2021). Virtual reality in health care: Bibliometric analysis. *JMIR Serious Games*, *9*(4), 1–19. https://doi.org/10.2196/32721

Powell, R., Ahmad, M., Gilbert, F. J., Brian, D., & Johnston, M. (2015). Improving magnetic resonance imaging (MRI) examinations: Development and evaluation of an intervention to reduce movement in scanners and facilitate scan completion. *British Journal of Health Psychology*, *20*(3), 449–465. https://doi.org/10.1111/bjhp.12132

Qian, K., Arichi, T., Price, A., Dall'Orso, S., Eden, J., Noh, Y., … Hajnal, J. V. (2021). An eye tracking based virtual reality system for use inside magnetic resonance imaging systems. *Scientific Reports*, *11*(1), 1–17. https://doi.org/10.1038/s41598-021-95634-y

Radomsky, A. S., Rachman, S., Thordarson, D. S., McIsaac, H. K., & Teachman, B. A. (2001). The claustrophobia questionnaire. *Journal of Anxiety Disorders*, *15*(4), 287–297. https://doi.org/10.1016/S0887-6185(01)00064-0

Rahani, V. K., Vard, A., & Najafi, M. (2018). Claustrophobia game: Design and development of a new virtual reality game for treatment of claustrophobia. *Journal of Medical Signals and Sensors*, *8*(4), 231–237. https://doi.org/10.4103/jmss.JMSS_27_18

Reiss, S. (1991). Expectancy model of fear, anxiety, and panic. *Clinical Psychology Review*, *11*(2), 141–153. https://doi.org/10.1016/0272-7358(91)90092-9

Richards, M. (2020). *Diagnostics: Recovery and Renewal*. Retrieved from https://www.england.nhs.uk/wp-content/uploads/2020/10/BM2025Pu-item-5-diagnostics-recovery-and-renewal.pdf

Rizzo, A. S., & Koenig, S. T. (2017). Is clinical virtual reality ready for primetime? *Neuropsychology*, *31*(8), 877–899. https://doi.org/10.1037/neu0000405

Sadigh, G., Applegate, K. E., & Saindane, A. M. (2017). Prevalence of unanticipated events associated with MRI examinations: A benchmark for MRI quality, safety, and patient experience. *Journal of the American College of Radiology*, *14*(6), 765–772. https://doi.org/10.1016/j.jacr.2017.01.043

Sanders, J. J., Caponigro, E., Ericson, J. D., Dubey, M., Duane, J. N., Orr, S. P., … Blanch-Hartigan, D. (2021). Virtual environments to study emotional responses to clinical communication: A scoping review. *Patient Education and Counseling*, *104*(12), 2922–2935. https://doi.org/10.1016/j.pec.2021.04.022

Sapkaroski, D., Mundy, M., & Dimmock, M. R. (2022). Immersive virtual reality simulated learning environment versus role-play for empathic clinical communication training. *Journal of Medical Radiation Sciences*, *69*(1), 56–65. https://doi.org/10.1002/jmrs.555

Spiegel, B. (2020). *VRx: How Virtual Therapeutics Will Revolutionaize Medicine*. New York: Hatchette Book Group.

Stunden, C., Stratton, K., Zakani, S., & Jacob, J. M. (2021). Comparing a virtual reality-based simulation app (VR-MRI) with a standard preparatory manual and child life program for improving success and reducing anxiety during pediatric medical imaging: Randomized clinical trial. *Journal of Medical Internet Research*, *23*(9). https://doi.org/10.2196/22942

The Royal College of Radiologists. (2017). *Magnetic Resonance Imaging (MRI) Equipment, Operations and Planning in the NHS Report from the Clinical Imaging Board. Royal College or Radiologists*. Retrieved from www.ipem.ac.ukwww.sor.org%0Awww.ipem.ac.uk%0Awww.sor.org

The Royal College of Radiologists. (2018). Sedation, analgesia and anaesthesia in the radiology department (second edition). *Clinical Radiology*. Retrieved from www.rcr.ac.uk

Thorpe, S., Salkovskis, P. M., & Dittner, A. (2008). Claustrophobia in MRI: The role of cognitions. *Magnetic Resonance Imaging*, 26(8), 1081–1088. https://doi.org/10.1016/j.mri .2008.01.022

Tischler, V., Calton, T., Williams, M., & Cheetham, A. (2008). Patient anxiety in magnetic resonance imaging centres: Is further intervention needed? *Radiography*, 14(3), 265–266. https://doi.org/10.1016/j.radi.2007.09.007

Tomaka, J., Blascovich, J., Kelsey, R. M., & Leitten, C. L. (1993). subjective, physiological and behavioural effects of threat and challenge appraisal. *Journal of Personality and Social Psychology*, 65(2), 248–260.

van Beek, E., Kuhl, C., Anzai, Y., Desmond, P., Ehman, R., Gong, Q., … Wang, M. (2019). Value of MRI in medicine: More than just another test? *J Magn Reson Imaging*, 49(7), e14–e25. https://doi.org/10.1002/jmri.26211.Value

Wardenaar, K., Lim, C., Al-Hamzawi, A., Alonso, J., Andrade, L., Benjet, C., … de Jonge, P. (2017). The cross-national epidemiology of specifc phobia in the world mental health surveys. *Psychol Med*, 47(10), 1744–1760. https://doi.org/10.1017/S0033291717000174 .The

Watson, R. E. (2015). Lessons learned from MRI safety events. *Current Radiology Reports*, 3(10), 1–7. https://doi.org/10.1007/s40134-015-0122-z

Weidman, E. K., Dean, K. E., Rivera, W., Loftus, M. L., Stokes, T. W., & Min, R. J. (2015). MRI safety: A report of current practice and advancements in patient preparation and screening. *Clinical Imaging*, 39(6), 935–937. https://doi.org/10.1016/j.clinimag.2015.09 .002

Wiederhold, B. K., & Bouchard, S. (2014). *Advanced in Virtual Reality and Anxiety Disorders*. London: Springer.

Wiederhold, B. K., & Wiederhold, M. D. (2000). Lessons learned from 600 virtual reality sessions. *Cyberpsychology and Behavior*, 3(3), 393–400. https://doi.org/10.1089 /10949310050078841

Wood, B. S., & McGlynn, F. D. (2000). Research on posttreatment return of claustrophobic fear, arousal, and avoidance using mock diagnostic imaging. *Behavior Modification*, 24(3), 379–394. https://doi.org/10.1177/0145445500243005

6 Ethical and Moral Considerations
Telerehabilitation

N Kirsch and G G Fluet

INTRODUCTION

Biomedical ethics is a relatively new branch of ethics but it is firmly rooted in the ancient foundation of ethical thought and reasoning, giving it strong credibility for the challenging ethical decisions facing clinicians in delivering healthcare services in contemporary society. Healthcare providers have a unique and special relationship with the individuals they are privileged to serve and this relationship requires strict attention to be paid to the understanding and application of ethical behavior that exceeds the standard in society. This moral obligation is based on trust and the fact that the healthcare provider is working with vulnerable populations, elevating the standard of care required.

Emerging technologies bring the promise of opportunities for patients and clinicians and the challenge of providing those opportunities safely and most effectively. Historically, there have always been new treatment techniques and methodologies challenging thought and practice but the context in which they are occurring from a societal perspective creates different dynamics. When new technologies become available, they are field-tested for their efficacy and safety. It is up to the professional to determine the appropriate application and that includes the professional judgment that is inherent in every clinical decision, which includes the questions: Is this the right person, the right place, and the right time for this particular treatment?

DECISION-MAKING FRAMEWORKS FOR ETHICAL ANALYSIS

The traditional frameworks for reasoning through an ethical situation provide guidance to the clinician when determining how to approach a question. The three frameworks, virtue, consequentialism, and principlism, permit a deep exploration into the ethical considerations of any decision.

Virtue ethics are described as the ethics of character asking the question: How should I be? In considering an action, the clinician may ask: Is what I am doing demonstrating the virtues/values that are expected of a healthcare professional?

DOI: 10.1201/9781003272786-6

In consequentialism, the clinician must consider the potential consequences of their proposed actions based on what may occur and who may be impacted as a result of that action. The "collateral damages" should be considered; this requires thinking beyond what is a potentially obvious impact on the patient and the clinician. What is the potential impact on other patients, family members, colleagues, society, and the reputation of the profession?

The duty of the healthcare provider is to behave in a manner consistent with the traditional ethical principles of beneficence, non-maleficence, autonomy, and justice. In principlism, a principle-based analysis also requires that the clinician considers a good balance between principles, not over-emphasizing one to the detriment of another. For example, in an attempt to be beneficent, the value of one treatment may be over-emphasized, but the patient may not be provided with adequate information or time to make a truly informed autonomous decision.

A discussion of the application of emerging technologies should begin with a broad look at the delivery of healthcare through the various modes of delivery commonly associated with telehealth and the special ethical considerations that are in place to help clinicians determine whether the delivery of safe and effective care is possible through telehealth. The core principles will be used for this exploration, followed by several cases in which a deeper ethical analysis, using the applicable decision-making framework, will be applied.

The balance of this chapter will review the ethical principles described above as they relate to emerging technologies in healthcare in general and telerehabilitation specifically. The chapter will conclude with two case studies demonstrating ethical analyses of two scenarios involving emerging healthcare technologies.

ETHICAL PRINCIPLES

BENEFICENCE

The ethical obligation to always do the most good possible in any given situation is at the heart of all clinician–patient interactions. Clinicians utilize accepted standards of care as the foundation for practicing in this fashion, adding new or alternative treatment approaches only when they will be equal to the standard of care, or provide benefits beyond it. New or alternative treatments are also attempted when aspects of the standard of care are difficult or impossible in specific clinical situations. The opportunity to provide rehabilitation interventions via telecommunications technology or any technology needs to be considered using these criteria each time this decision is made. Key questions are related to the appropriateness for the patient, the patient's condition, the stage of this condition, and whether this is the ultimate treatment venue, hence "right person, right time, right place." In addition, the available evidence to support the intervention and the competence of the provider need to be considered to decide on the best interests of the patient.

Provider Competence

Rehabilitation professionals enjoy the privilege of a broad and varied scope of clinical practice which results in the obligation that clinicians evaluate their own level of competence in clinical practice and only engage patients in clinical situations when it is in the patient's best interest. For it to be ethical for them to see patients in this setting, clinicians must be confident that they can collect a sufficient amount of quality data via telerehabilitation platforms to make effective clinical decisions. In addition, they must be able to utilize telerehabilitation software and hardware well enough that their use of the technology does not detract from the ability of patients to receive quality services. This level of skill needs to extend to the ability to trouble-shoot simple and typical problems encountered by patients using the telerehabilita-tion system employed by their clinic. Finally, it is important to note that this level of technical skill does not necessarily translate into an appropriate level of clinical skill. Providers must <u>also</u> have the requisite clinical knowledge and experience to handle a patient's diagnosis and clinical presentation effectively.

EBP

There is a growing evidence base related to telerehabilitation that should clarify the decision to work with patients in a telerehabilitation setting versus encouraging a patient to seek in-person rehabilitation treatments. Major systematic reviews support the view that telerehabilitation is as effective as or more effective than in-person rehabilitation for chronic obstructive pulmonary disease (Cox et al., 2021), stroke (Laver et al., 2020), and lower extremity joint replacement (Jansson et al., 2022) Similar reviews of smaller bases of lower-quality evidence suggest that telereha-bilitation may be effective for improving the physical function of persons with mul-tiple sclerosis (Rintala et al., 2018; Khan, 2015), shoulder pain (Gava et al., 2022), Parkinson's disease (Vellata et al., 2021), osteoarthritis (Pietzrak et al., 2013), and cardiovascular disease (Chanet al., 2016). In addition, a strong, but less-than-specific review of unspecified musculoskeletal disorders suggested that synchronous telere-habilitation was as effective as or more effective than in-person rehabilitation for decreasing pain and increasing function in this family of diagnoses (Cottrell et al., 2017). Therapists working with clients with these diagnoses can use this evidence to support decisions for these clients in a telerehabilitation environment. An important counterpoint to these findings are studies identifying manual therapy techniques as key interventions in the management of mechanical neck disorders (Gross et al., 2002), and plantar fasciitis (Fraser et al., 2018). The ethical principle of beneficence would dictate that the best plan of care possible for these patients includes manual therapy, which is difficult to impossible to perform effectively via telerehabilitation, therefore clearly pointing to the fact that these interventions should be implemented in person.

Stage of Injury

Acute injuries, especially those that cause a client pain, must be handled differently from chronic conditions. Many of the passive modalities that are available in-clinic, including hands-on treatments, such as massage, can provide a patient comfort during the acute phase of an injury (Ebadi et al., 2020; George et al., 2021). The evidence base supporting the contribution that many of these treatment techniques make to recovery is questionable, but a reduction in suffering is a beneficent act.

When discussing the relative benefits of in-person versus telerehabilitation treatments, a provider is obligated to discuss the availability of treatments that might make a patient more comfortable in the short term but that will not necessarily provide patients with a better recovery in the longer term.

Patients Appropriate for Telerehabilitation

Telerehabilitation interventions tend to place an increased burden on clients and caregivers, as compared with clinic visits following the need to transport themselves to the clinic has been accomplished. To be most effective, proactive participation is necessary to set up and troubleshoot the software used for the exchange at the patient end. In addition, clients will need to manage multiple electronic devices and deal with connection issues on their end. Many clients will need to rearrange the area in their home where treatments will occur to accommodate the requirements to be prone, supine, side-lying, sitting, standing or walking, while sufficiently on-camera for the clinician to see. Accommodation for sufficient quiet and privacy during sessions will need to be arranged. Clients will need to dress themselves for rehabilitation and disrobe and dress themselves again (including shoes) without assistance or they will need to solicit the assistance of a family member or caregiver for each treatment session. It is important to note that these issues need to be managed for every single session of a plan of care. All patients may not have the psychological/emotional wherewithal to meet these demands, so establishing that a patient and a necessary caregiver, if required, is able to handle these details consistently on an ongoing basis is required for such a plan of care to be successful and critical to determine that a telerehabilitation-based plan of care is the best option for a patient.

Home Environment Appropriate for Telerehabilitation

Camera-based telerehabilitation requires a fair amount of space, especially for patients who will perform standing or walking activities during a treatment. Patients living in smaller spaces may not be able to perform important aspects of a plan of care that includes these more challenging postures. If the end of a plan of care should require standing or walking and that cannot be accomplished safely and effectively, a plan of care that includes a transition to in-clinic visits or 100% in-clinic visits from beginning to end should be presented as options to patients and caregivers.

Access to appropriate technology for interaction in a synchronous telerehabilitation environment is one of the obvious issues to consider when assessing a client for a telerehabilitation-based plan of care. Consistent internet access, a router with sufficient speed, and a computer capable of running the telerehabilitation software are necessary. Identifying patients with access to this technology is necessary before an episode of care is initiated via synchronous telerehabilitation.

Caregivers that will be available for an entire plan of care and are physically able to assist as necessary for all activities likely to be part of an episode of care are necessary when a patient is unable to perform these tasks for themselves. Consistent availability is critical to ascertain as well.

NON-MALEFICENCE

The concept of doing no harm carries just as much weight during digital and telerehabilitation-based encounters with patients as it does during in-person interactions. Telerehabilitation places limitations on the screening process, making it a challenging aspect of the plan of care to conduct in an ethical fashion for all patients. Screening for cardiovascular issues is more difficult in a telerehabilitation setting because blood pressure and pulse oximetry equipment designed for home use tend to be less accurate than clinical-grade equipment used by a healthcare practitioner (Shimbo et al., 2020; Harskamp et al., 2021), particularly in patients with dark skin (Kyriacou et al., 2022). Skin cancer screening can be less effective if clinicians are not trained in remote skin cancer screening (Macbeth et al., 2015). Skilled palpation is also an important examination modality that is lost during telerehabilitation-based examinations; screening for masses in patients with soft tissue abnormalities and pain will be less effective in the unskilled hands of patients or caregivers. Patients with known cardiovascular disease and personal or family histories of cancer may need to be examined in-person for a telerehabilitation-based plan of care to be considered safe.

Balance assessment is another aspect of rehabilitation examination that must be modified to be feasible and safe in a telerehabilitation setting. Loss of balance is the ultimate goal of balance testing because it is the most valid indicator that an activity or posture exceeds a patient's abilities. This said, testing that elicits a loss of balance must be guarded by a reliable caregiver and every loss of balance guarded by an untrained non-professional increases the risk of a fall with injury (to the patient, caregiver, or both). This results in an ethical challenge. Sufficiently challenging tasks are necessary for accurate diagnosis and effective management of balance disorders, which are necessary to maintain or improve a patient's safety, but deliberately causing the potential for a client to fall and providing inadequate safety protection under sufficiently challenging examinations is inherently risky.

In addition to screening concerns and examination issues, medical stability needs to be considered when assessing a patient's appropriateness for telerehabilitation. Patients that are sufficiently stable to be at home are not necessarily stable enough for rehabilitation interventions. This puts a clinician in the position of balancing choosing an exercise intensity that will be effective and an exercise intensity that will be safe, considering difficulties in monitoring a patient's response to exercise and the lack of clinicians to attend to the patient in an emergency. A level of instability that precludes a clinician from achieving the greatest good possible (beneficence) while avoiding the possibility of doing harm (non-maleficence) constitutes an ethical distress. In this situation, all possibilities for in-person or in-clinic rehabilitation must be exhausted before the choice to provide telerehabilitation services could be considered ethical.

Autonomy

Respecting the ethical principle of autonomy obliges healthcare professionals to share case-specific information and options necessary for patients to exercise self-determination as it relates to their care (Varkey, 2021). Central to respecting autonomy is the ethical and legal construct of informed consent. Providers offering a telerehabilitation option to current or prospective patients and their families/caregivers must provide a thorough review of the relative risks and limitations associated with remote supervision of patients including: 1) Decreased information available to the clinician (lower resolution 2-D visual information only, inability to palpate, less auditory information, no smell), and the impact this has on patient safety, i.e., less effective screening for medical issues; 2) Increased reliance on patients and caregivers for vital sign monitoring and guarding; 3) Access to less equipment; 4) Lack of access to hands-on interventions; 5) Decreased access to emergency medical services (particularly when compared with hospital-based clinics); 6) Decreased opportunities for peer support. This is a particularly nuanced conversation as the relative negative impacts of these issues need to be balanced in relation to patient demographics, nature of condition, stage of recovery, presence of comorbid conditions, and the feasibility of in-clinic treatments. Alternatives to telerehabilitation need to be presented as well. The option to receive home care services for homebound clients needs to be presented, in addition to the availability of home-based outpatient services for patients who are not homebound.

In addition to clinical issues, telerehabilitation presents data privacy issues beyond those posed by modern in-person rehabilitation. Although all encounters should respect patient privacy and confidentiality, the increase in telephone- and web-based interactions increases the opportunities for data to be stolen. Steps taken to ensure privacy need to be described to the client, but the increased risks associated with telephone or online interactions need to be shared with the client as well. In addition, if data are collected and managed by third-party companies, every safeguard must be taken to prevent a data breach, and the use of a third party must be shared with the clinician.

Potential, perceived, or real conflicts of interest need to be divulged to clients as well. Clinician investment in or partnership with companies providing telerehabilitation services need to be disclosed. Telerehabilitation patients must be informed if they or their data are being used in healthcare research, particularly if telerehabilitation is central to the research study.

Costs are another factor that needs to be discussed. Clinicians need to be aware if telerehabilitation patients are being charged technology fees, and technology charges need to be shared with clients. It is also necessary for clinicians to inform patients if they are charging them or third-party payors for telephone consultations or for viewing text- or telecommunications-based movies related to their care between sessions. Differences in cancellation policies, especially cancellations caused by technology issues at either end, need to be clearly explained to clients, and differences between telerehabilitation and in-person cancellations need to be made clear. Finally, liability issues need to be discussed with clients as they relate to injuries that occur as a result of conditions in their own homes.

It is important to note that, when the differences between in-person and telerehabilitation have been responsibly relayed to the patient, the patient still needs to be able to decide to go with the delivery method that they are comfortable with, and this decision needs to be honored even when logistical considerations favor telerehabilitation for clinicians and caregivers. Autonomy needs to continue after a plan of care has been initiated as well. If a patient decides that telerehabilitation "isn't for them" after a few sessions, clinicians treating them are obligated to honor these wishes by transitioning their patient to in-person sessions or referring them to clinicians who provide these services.

JUSTICE

Telerehabilitation addresses two issues that can serve to expand access to care, increasing the just distribution of healthcare resources. Both the urban poor and the rural poor are groups with reduced access to many types of care including rehabilitation. Potential clients unable to access care due to the cost of travel back and forth from a clinic and the lack of free transportation services in their area are much more likely to receive quality care if telerehabilitation services are offered. These benefits also extend to the urban poor living in older buildings with structural barriers that make regular trips to a clinic difficult to impossible as well. These benefits are accompanied by potential abuses. The urban poor often receive care via the home healthcare system. Home healthcare companies and personnel often take on risks when operating in high crime areas. Although offering services to patients in the home via telerehabilitation would seem like an excellent solution to this problem, the solution becomes less than fair when it involves patients who are less appropriate for telerehabilitation services than they are for in-person home treatments. Justice would require that patients in high- and low-crime areas receive similar access to the most appropriate care. The efficiency of telerehabilitation also presents opportunities for abuse. Telerehabilitation requires little to no clinic space, none of the equipment which is necessary in an in-person clinic, and less facility liability insurance. In addition, telerehabilitation is associated with decreased cancellations, making telerehabilitation patients less costly to treat than in-clinic patients. This could lead to clinics funneling patients from lower-margin payors, like Medicaid or other Affordable Care Act-associated plans, toward telerehabilitation treatments in an attempt to maximize the return on limited reimbursement. Again, justice dictates that high- and low-reimbursement patients receive the same care, given the same clinical situations.

Home environment and computer literacy interact with the ethical principle of justice as well. Most of the space and privacy limitations and technology access situations, as well as the computer literacy issues described in this chapter, occur in greater rates in economically under-resourced communities. Healthcare organizations and providers profiting in the telerehabilitation market carry ethical burdens to address these disparities. Urban medical centers could support community-based telerehabilitation satellite clinics in urban areas. Rehabilitation clinics in rural areas could offer technology-lending programs to patients living in remote areas. Companies producing telerehabilitation systems could develop systems that operate on low-cost hardware, such as Chromebooks, and offer lower-cost multiple-user subscriptions to clinics serving poor communities. Developers could produce interfaces that require lower levels of computer literacy to operate. It is important to note that, apart from the hospital systems and clinics, these parties are for-profit entities without the same ethical responsibilities and accountability that healthcare providers carry. The responsibility for choosing products that optimize access to telerehabilitation for under-resourced patients and designing implementation budgets that include strategies targeting access for these clients falls on clinicians and clinician/managers. There is a developing literature examining technology used to provide rehabilitation services in a more just fashion. Some notable articles include studies of traditional telecommunications (Kim and Zuckerman, 2019), mobile technology applications (van Veen et al., 2019), and social media (Gibbons et al., 2011) applications.

MODELS OF TELEREHABILITATION

Pure Telerehabilitation (PTR) All sessions are delivered via telerehabilitation. Similar combinations of initial evaluations, treatment sessions, check-ins, follow-ups, re-evaluations, and discharge evaluations can occur.

Hybrid Sessions These are a combination of telerehabilitation and in-person rehabilitation. Typically, initial evaluations are performed in person. Subsequent treatment sessions can be a combination of in-person visits or telerehabilitation visits. Check-ins or follow-ups can be in person but are often delivered via telerehabilitation. Re-evaluations and discharge evaluations are typically performed in person. Episodes of care can be extended with check-ins or follow-ups when regularly scheduled sessions have been put on hold or discontinued. The flexibility that a hybrid model affords allows clinicians and clients to make the most effective decisions based on clinical considerations and personal/logistical considerations. This model, when judiciously used, can decrease the likelihood of ethical issues.

Synchronous Sessions These can be used in hybrid and PTR models. These sessions involve communication between the provider and patient that occurs in real time, with the potential to exchange information in both directions at the same time. These sessions can occur via telephone or internet. They can involve voice and visual communications, as well as the real-time exchange of digital, sensor-based data (wearable sensors or cameras).

Asynchronous Sessions These can be used in hybrid and PTR models as well. These sessions involve providers reviewing voice, photo, video, or sensor data provided to them by patients or caregivers and using clinical knowledge or judgment to respond to these data appropriately. The use of sensor data, particularly camera-based and inertial measurement unit (IMU)-based data, poses issues when used in an asynchronous setting. A majority of the research done examining the reliability and validity of these approaches to measuring and evaluating human movement has been done in lab settings with healthy subjects. The reliability and validity of these measures suffers when clinical populations use these technologies in clinical and home settings (Tack, 2019). This mismatch requires an active effort on the clinician's part to rectify the data they use to make clinical decisions and also to help the patient interpret feedback that the system may provide, based on data collected passively by sensors. Other issues, the receipt of, review of, and responses to these data by the clinician are often separated from each other by time. Time lags are an important consideration when safety (non-maleficence) and effectiveness (beneficence) are considered. Standards for asynchronous monitoring have not been established but client characteristics, such as the nature of the condition and the medical condition of the patient, need to be considered when choosing the time frame over which a patient needs to be monitored and given direction/feedback.

Telecommunication-based Telerehabilitation This is limited to voices being communicated in both directions via land-based telephone lines. It provides clinicians and clients with the least amount of information (audio only), increasing concerns related to safety and effectiveness.

Mobile-based Telerehabilitation This occurs by utilizing cell phone technology, involving voice, text, photo, video, and mobile phone-based sensor data, which can include inertial measurement unit data and satellite (global position system, GPS) data. GPS data offers providers with rich information related to their clients' participation in society, that goes well beyond the rough estimates of activity that inertial measurement units provide. Sharing this type of information can make a patient's own individual care more effective and also allows clinicians to examine their practices across a group of patients, improving care for everyone. Informing patients that the GPS features "track their every move" when these features are utilized is a key feature of the informed consent process.

Web-conferencing-based telerehabilitation This uses commercial conferencing services, such as Webex or Zoom, to transmit voice, text, photo, and video. These systems are not designed specifically for healthcare. Privacy concerns begin to arise when these systems are used. Many services offer upgraded packages that offer more security/privacy.

Telehealth systems These are server/software packages specifically designed to provide a wide variety of healthcare services. Some of these systems are embedded in electronic medical record systems. Typically, they are compliant for privacy concerns and data transmission/storage.

Telerehabilitation systems These are server/software packages specifically designed to provide rehabilitation services. User interfaces often feature rehabilitation specific-fields that facilitate user experience. These systems may also include rehabilitation-specific features such as electrogoniometry or dynamometer interfaces.

Digital health is a fast-growing sector of the healthcare industry that leverages various technology-based health delivery methods. Wearable devices, telecommunications, and mobile communications technology are utilized to provide rehabilitation and health promotion services by several digital health companies. Some digital health companies have focused on traditional, synchronous telerehabilitation approaches whereas others utilize user-centered and data-driven programs that de-emphasize traditional patient–clinician roles. Clinicians need to be conscious of the non-traditional roles that these services afford when becoming affiliated with digital health providers. Clear descriptions of the roles that clinicians have with digital health clients are the ultimate responsibility of the affiliated clinicians. If a client will actually be "supervised" during rehabilitation programs by an algorithm based on their movement data, a bot using language recognition, or a health coach without formal training (even if the clinician helped design these services), this needs to be clear, and the nature of actual individual supervision provided by clinicians must be clear as well. If these conditions are not met, the clinician has not provided a level of veracity necessary for the client to act autonomously. One of the selling points often associated with digital health services are cost savings associated with their use. This can set up potential justice issues when employees of a company are encouraged to use more cost-efficient digital health services that might be less effective than in-person, professionally supervised healthcare for specific clients and specific conditions. Awareness of breaches of term and title protection must also be considered with digital health, for example, calling a session not provided by a physical therapist or a physical therapist assistant "physical therapy" would be illegal in most jurisdictions.

Many digital health platforms incorporate various forms of artificial intelligence (AI) in their management of patients. In digital rehabilitation settings, demographic data, patient reports of symptoms, past and family medical history, and data from wearable sensors are utilized to develop movement dysfunction diagnoses for patients with neuromuscular and musculoskeletal disorders. Algorithms analyze the data from large groups of patients with known diagnoses to identify data clusters that can be used to predict the presence or absence of movement dysfunctions in new clients that are yet to be diagnosed, and then recommend intervention. These diagnoses and the treatment plans generated are weakened by two issues: first, the fact that most of the observational data fed into these machine learning algorithms are subject to confounding and selection bias, and second, the fact that predictions based on modeling of observational data cannot logically establish causation, which requires the formation and testing of alternative hypotheses. Currently, AI-based rehabilitation still requires a machine/human (with domain-specific expertise) partnership for effective diagnosis and intervention (Prosperi et al., 2020). A jump to the use of unsupervised AI at this level of development in rehabilitation would therefore constitute practicing below the standard of care (violating the principle of beneficence) at a higher risk of injury to the client should a misdiagnosis occur (violating the principle of non-maleficence).

The limitations of AI also provide a call to action to clinicians. Precision/data-driven care holds the promise of substantially improved clinical outcomes and digital/telerehabilitation holds the promise of substantial improvements in affordable access to care. It will be necessary for clinicians to partner with data scientists to improve the quality of data collected in routine clinical practice and translate clinical reasoning in a fashion that can be used to take AI-based care beyond predictive modeling. Eventually, it will also be necessary for clinicians to abrogate certain aspects of clinical practice to machines that can perform these functions at a higher degree of efficiency and accuracy than they are able to, and focus on the aspects of care that require human interaction.

Artificial intelligence is rich with potential but offers new ethical challenges for all those in a decision-making capacity regarding introducing new technology into the healthcare delivery system. Established ethical principles in healthcare, autonomy, respect for persons, beneficience, caring, and the guiding principle of "do no harm" (non-maleficence) are truly at play when introducing the use of AI into the decision-making process. Patients must have information provided to them in a way in which they can understand it to make an informed decision, weighing the risk-benefit ratio of any proposed intervention.

Concerns arise that new stakeholders have entered the healthcare space, including technology giants and startups in digital technology. Data quality is often questionable, which is compounded by a lack of regulatory oversight or accepted standards in this new frontier of healthcare. It appears that new technologies are finding themselves in both the research and clinical practice worlds without traditional vetting (Nebeker et al., 2019). Stakeholders must be encouraged to recognize that the tools they wish to introduce are often targeting vulnerable populations, increasing the ethical stakes in decision-making. The digital health decision-making framework introduced by (Pagoto and Nebeker, 2019; Nebeker et al., 2018) was developed to help make sound decisions in digital technology research and is applicable to the use of artificial intelligence in healthcare. The framework consists of five domains: 1) Participant privacy, 2) Risks and benefits, 3) Access and usability, 4) Data management and 5) Ethical principles (Nebeker et al., 2020). It is important to note that the five domains are designed to have a relationship between them that intersects. Consistent use of tools such as this at each stage of development, and implementation of new rehabilitation technologies will do much to insure that all parties are acting in the best interest of patients.

CASE 1: DIGITAL HEALTH DILEMMA

Joe is a 56-year-old English Composition teacher at three community colleges in his area. Almost all of his time is spent driving between his home and the colleges he works at and grading essays. Joe's back started bothering him about six months ago and has gotten stiffer each month. He has trouble finding a comfortable driving position and cannot stand up after sitting for more than 15 minutes, which makes grading essays take even longer. A few weeks ago, Joe received a flyer in the mail about a "digital health platform" that members of his employee health plan could access free of charge. One of the services offered was "exercise therapy." Joe knew he needed to do something about his back but could not afford the copay for physical therapy, and a gym membership or personal trainer was out of the question. After signing up, Joe answered about 100 online questions about his health, his back pain, and his goals. He also got texts from Violet, his physical therapist, and a health coach named Tim that would be working with him. He was instructed to send messages to the number provided to communicate with his team. Joe googled Violet's name and was excited to see that she was a physical therapist in a neighboring city.

A few days later, he got a tablet and some sensors in the mail. Joe tried the first day of exercise therapy as soon as he opened the box. He strapped on the sensors and did a few tests that made him move in various directions with the sensors measuring how far he bent. The next day, his tablet guided him through six exercises, with the sensors measuring how far he moved. A few made him move in similar directions to the tests from the day before and two were new. One of the new stretches was impossible but the other seemed to work well. After he completed his exercises, Joe typed into a feedback box that he was unable to do the stretch. Coach Tim texted him later that day and emailed him a set of alternative stretches a few minutes later. He instructed Joe to substitute one of the alternatives to cue him the next time he did his exercise therapy. Joe tried all the new stretches and chose the one that worked best.

A couple of days later, Joe struggled with the sensors. The graphics on his tablet did not seem to reflect how far he was bending and kept telling him to bend further. The next time he exercised, he chose the no-sensor option and completed his exercises. Coach Tim texted him later that day, saying that it was OK, that the sensors were used for motivation, not tracking.

About a week later, Joe noticed some pain in his buttock as he performed a seated side-bending exercise. The pain lasted about an hour and then wore off, but it bothered him the next morning again when he woke up. Joe texted the number he was given, informing his team of his new symptoms. Coach Tim responded that pain did not equate to harm and that his pain should decrease as his strength and flexibility improved.

Violet graduated from an excellent Physical Therapy education program three years ago. She is employed at a busy outpatient clinic and loves her job, but, up until last year, she was unable to move out of her parents' home because she was unable to afford rent while she paid off her student loans. Last year, Violet took a job with GTR Health, a rapidly expanding company that offered digital health services, targeting persons with musculoskeletal pain. Her assignment would be to assist with the programs of up to 100 people with musculoskeletal pain. Programs were provided to clients via an algorithm that chose a set of exercises based on the region of their body that caused them pain and the programs were modified based on the number of sessions that clients exercised each week and their reported pain levels were entered on the app. An administrative assistant and a coach would track and provide support to her clients daily and alert her if the clients reported issues that required her expertise. Violet would need to log into the system every 48 hours (about two days) and was asked to message at least five clients per log in. This seemed reasonable to Violet, and she was encouraged to see that all her clients were from her home state, which eliminated some potential compliance issues with her state. Although most clients dropped off after a few weeks, the clients that stuck with their programs seemed to be reporting improvements. The money Violet received covered her student loan payment nicely and allowed her to move into a townhouse with a friend.

About six months into her employment with GTR Health, Violet noticed some issues with the way her clients were being handled. Occasionally, her client load increased beyond her limit and sometimes clients were employees of companies in her state that lived in an adjoining state. In addition, Violet noticed that coaches were giving advice to clients that were describing problems with their exercise program that should have been referred to her. Finally, several of the coaches referred to Violet as "your therapist" when communicating with clients. Violet attempted to communicate directly with the coaches, that, while she was a therapist, she could not be considered a client's therapist if they were not being managed by her or a licensed physical therapist practicing under her supervision 100%, but some of the coaches did not seem to understand the distinction. Violet liked the premise of digital health and how accessible it is to patients but there were some things that she was feeling rather uncomfortable about, but she was not sure if the discomfort outweighed that student loan paydown and moving out of her parents' basement.

CASE 1 ETHICAL ANALYSIS

Using the three frameworks to do an ethical analysis, Violet first considered the **virtues** expected of her as a healthcare provider. Were her reasons for providing digital healthcare consistent with the core values of her profession? While she could convince herself that what she was doing was out of compassion, providing care that the patient may not have had access to otherwise, the virtues of altruism and collaboration were more difficult to see since Joe was not improving and she did not recommend a perhaps more appropriate level of care. One could also question if she was truly fulfilling her professional duty or demonstrating excellence. One might also consider the potential for a conflict of interest as her choice to continue his care with a less-than-effective means stood to benefit her loan repayment more than his return to pain-free function.

Next, Violet turned to a **principle-based assessment**: were the interventions truly demonstrating "care," were they addressing his specific problem as a healthcare provider is required to do? Was her care potentially maleficent, causing harm as she did not refer him when he was not making progress and if he were in her care in another setting would he have been treated in a similar manner? This is unlikely as, in an actual physical therapy setting, he would have had more options for care. Was she honest and truthful about his care and did she provide him with options and alternatives? Finally, examining the situation from a **consequence-based approach**, what was the outcome for Joe? If Joe thought this is the best he could get with what he considered to be physical therapy, would he try any other type of physical therapy in the future? What is his impression of the profession? Was the money well spent by the third party payor? Did Violet provide care she was proud of as a professional? The technology was valuable, but did she make the best and most effective use of it?

CASE 2. ARTIFICIAL INTELLIGENCE: JUST HOW SMART ARE WE?

Marcus was always intrigued by how he could integrate his fascination with artificial intelligence (AI) and his commitment to providing the best care possible for his patients. Although he did not think his patients were shortchanged by his treatment decisions, he could not help but think that some of the advanced decision-making capabilities of AI could put another really powerful tool in the treatment toolbox he was always adding to. Whereas his colleagues showed an interest in the topic, none of them seemed to share his passion for trying to integrate AI into treatment … they all seemed to think it was something kind of interesting they may see some use of in the future. Marcus was convinced that the future was already here and, priding himself on being an early adopter, he did not want to be standing on the sidelines as AI left the theoretical and became mainstream. Marcus was approached by a small startup company who knew of his work with children with cerebral palsy who exhibited upper extremity use-neglect patterns. They developed an app with which they claimed children would use the neglected extremity at least 50% more often than any other therapeutic intervention, to increase use. Marcus was intrigued, of course, and the app was fun to use so he was pretty sure the children would be drawn to its use.

The company, a small startup, made no claims of providing actual treatment or providing any type of diagnosis, so US Food and Drug Administration approval was not required. There were no publications documenting that the product was safe or valid or reliable; the only formally acknowledged data involved use and children were interested in using the product. The terms and conditions were not particularly clearly written and all that was required to start using it was for the app to be downloaded to a tablet, or other smart device. The algorithm controlling the app was proprietary and therefore there was no way that an outside entity could determine the decision-making process that went into its workings. Data collected through the app about the children using it was the property of the company. Marcus could not help but be very interested in being involved in some level on the introduction of this app as a possible adjunct to his treatment…on the other hand, everything that he learned regarding evidence-based practice was not demonstrated in what the company was presenting to him. He didn't want to deny his patients every opportunity to improve, yet he knew he had a deep personal and professional commitment to be a responsible clinician, protecting the health and safety of his young patients. The decision on whether to work with the startup was not a simple one and he was confident that an ethical analysis of the options was in order.

Case 2 Ethical Analysis

Marcus began his analysis from a **virtues** perspective. Although he could clearly demonstrate that his interest in engaging children and providing meaningful treatment options for them was consistent with the important virtue of altruism, he also recognized that using a treatment technique that was not well vetted and had little evidence to support it was not consistent with the virtues of accountability, excellence, or professional duty. Marcus continued his analysis from a **principle-based approach**, beginning with the respect for individuals that is demonstrated through preserving autonomy through informed consent. Here, he struggled as the elements of informed consent that should form the framework for the patient (or family) to be able to make a reasoned decision are not present. He was convinced that engaging his young patient in a fun therapeutic activity would be beneficial to his patient, but he also had the foresight to look ahead and consider that what appeared to be beneficial could actually be harmful physically and emotionally to the child if it either did not work or caused some actual physical harm. Marcus recognized that this new technology had limited reach across the patient community and may be another example in which the principle of justice was not being upheld and we were not addressing real and perceived health disparities. Finally, Marcus considered the use of the app from a **consequence-based perspective**. What is the impact of this technology on the patient's care? Though it is fun and engaging, it is taking valuable time that may be put to better use by a trained clinician. Is the risk-benefit ratio sufficient to justify the use of the app? Are there harms to the profession, in terms of the loss of trust in the judgment of clinicians when technology is initiated that may not have the sound backing of science? With all of these considerations Marcus was concerned that his personal interest and fascination with artificial intelligence may be at the root of a conflict of interest in which he wants to initiate these "cool new things" but they may not be in the patient's best interest to continue to pursue care that has not been well established. Marcus is considering his options and his ethical obligations appear to lean toward more traditional care while continuing to learn about how to best incorporate emerging technology safely and effectively into contemporary practice.

CONCLUSION

Emerging technologies, similar to any of the changes in the delivery of healthcare over the centuries have raised questions of value, efficacy, safety, and necessity. The explosion of growth in technology and the costs related to that growth increase the complexity of the decision-making process as clinicians deal with the choices of whether or not to introduce a new technology into healthcare. Ethical frameworks help to organize the decision-making process. Is the treatment technique of benefit to the patient? Is the clinician implementing a treatment protocol because it is best for the patient or because it is best for the clinician? What are the virtues demonstrated by the clinician in making the decision they are making? Which principles is the clinician considering? Has the patient been informed appropriately? Will using a technique or not using it potentially cause harm? What about the principle of justice? Does the use of technology further increase healthcare disparities? Can technology decrease disparities? Finally, what are the consequences of our actions as healthcare providers? Have we taken the opportunity to consider the ramifications of our actions, taken the time to look not just straight ahead but around the corner to see what may occur? Are we aware of the ripple effect of what we are doing? What is the impact on patients, caregivers/family, colleagues, other clinicians, third-party payors, our profession, our academic and regulatory communities? All of these ethical constructs are critical as we consider how we best utilize emerging technologies. "Just because we can … doesn't mean we should." Our patients deserve critical assessment based on sound fact when we introduce anything that has the potential to have a therapeutic impact.

REFERENCES

Chan, C., et al., Exercise telemonitoring and telerehabilitation compared with traditional cardiac and pulmonary rehabilitation: A systematic review and meta-analysis. *Physiotherapy Canada*, 2016. 68(3): 242–251.

Cottrell, M.A., et al., Real-time telerehabilitation for the treatment of musculoskeletal conditions is effective and comparable to standard practice: A systematic review and meta-analysis. *Clinical Rehabilitation*, 2017. 31(5): 625–638.

Cox, N.S., Dal Corso, S., Hansen, H., McDonald, C.F., Hill, C.J., Zanaboni, P., Alison, J.A., O'Halloran, P., Macdonald, H., & Holland, A.E. Telerehabilitation for chronic respiratory disease *Cochrane Database of Systematic Reviews* 2021, Issue 1. Art. No.: CD013040.

Ebadi, S., et al., Therapeutic ultrasound for chronic low back pain. *Cochrane Database of Systematic Reviews*, 2020(7).

Fraser, J.J., et al., Does manual therapy improve pain and function in patients with plantar fasciitis? A systematic review. *Journal of Manual & Manipulative Therapy*, 2018. 26(2): 55–65.

Gava, V., et al., Effectiveness of physical therapy given by telerehabilitation on pain and disability of individuals with shoulder pain: A systematic review. *Clinical Rehabilitation*, 2022. 36(6): 715–725.

George, S.Z., Fritz. J.M., Silfies, S.P., Schneider, M.J., Beneciuk, J.M., Lentz, T.A., Gilliam, J.R., Hendren, S., & Norman, K.S. Interventions for the Management of Acute and Chronic Low Back Pain: Revision 2021. *J Orthop Sports Phys Ther.* 2021. 51(11): CPG1-CPG60.

Gibbons, M.C., et al., Exploring the potential of web 2.0 to address health disparities. *Journal of Health Communication*, 2011. 16(suppl1): 77–89.

Gross, A.R., et al., Clinical practice guideline on the use of manipulation or mobilization in the treatment of adults with mechanical neck disorders. *Manual Therapy*, 2002. 7(4): 193–205.

Harskamp, R.E., et al., Performance of popular pulse oximeters compared with simultaneous arterial oxygen saturation or clinical-grade pulse oximetry: A cross-sectional validation study in intensive care patients. *BMJ Open Respiratory Research*, 2021. 8(1): p. e000939.

Jansson, M.M., et al., The effects and safety of telerehabilitation in patients with lower-limb joint replacement: A systematic review and narrative synthesis. *Journal of Telemedicine and Telecare*, 2022. 28(2): 96–114.

Khan, F., Amatya, B., Kesselring, J., & Galea, M. Telerehabilitation for persons with multiple sclerosis. *Cochrane Database of Systematic Reviews*, 2015, Issue 4. Art. No.: CD010508.

Kim, T. and J.E. Zuckerman, Realizing the potential of telemedicine in global health. *Journal of Global Health*, 2019. 9(2): 020307.

Kyriacou, P.A., et al., *Inaccuracy of pulse oximetry with dark skin pigmentation: Clinical implications and need for improvement.* 2022. Elsevier.

Laver, K.E., Adey-Wakeling. Z., Crotty, M., Lannin, N.A., George, S., & Sherrington, C. Telerehabilitation services for stroke. *Cochrane Database of Systematic Reviews* 2020, Issue 1. Art. No.: CD010255.

Macbeth, F., et al., Melanoma: Summary of NICE guidance. *BMJ*, 2015. 351.

Nebeker, C., R.J. Bartlett Ellis, and J. Torous, CORE tools. *Digital health decision making framework and checklist designed for researchers.* 2018. https://recode.health/dmchecklist. Accessed: August 5, 2022.

Nebeker, C., J. Torous, and R.J. Bartlett Ellis, Building the case for actionable ethics in digital health research supported by artificial intelligence. *BMC Medicine*, 2019. 17(1): 1–7.

Nebeker, C., R.J. Bartlett Ellis, and J. Torous, Development of a decision-making checklist tool to support technology selection in digital health research. *Translational Behavioral Medicine*, 2020. 10(4): 1004–1015.

Pagoto, S. and C. Nebeker, How scientists can take the lead in establishing ethical practices for social media research. *Journal of the American Medical Informatics Association*, 2019. 26(4): 311–313.

Pietrzak, E., et al., Self-management and rehabilitation in osteoarthritis: Is there a place for internet-based interventions? *Telemedicine and E-health*, 2013. 19(10): 800–805.

Prosperi, M., et al., Causal inference and counterfactual prediction in machine learning for actionable healthcare. *Nature Machine Intelligence*, 2020. 2(7): 369–375.

Rintala, A., et al., Effectiveness of technology-based distance physical rehabilitation interventions on physical activity and walking in multiple sclerosis: A systematic review and meta-analysis of randomized controlled trials. *Disability and Rehabilitation*, 2018. 40(4): 373–387.

Shimbo, D., et al., Self-measured blood pressure monitoring at home: A joint policy statement from the American Heart Association and American Medical Association. *Circulation*, 2020. 142(4): p. e42–e63.

Tack, C., Artificial intelligence and machine learningl applications in musculoskeletal physiotherapy. *Musculoskeletal Science and Practice*, 2019. 39: 164–169.

van Veen, T., et al., Potential of mobile health technology to reduce health disparities in underserved communities. *Western Journal of Emergency Medicine*, 2019. 20(5): 799–802.

Varkey, B., Principles of clinical ethics and their application to practice. *Medical Principles and Practice*, 2021. 30(1): 17–28.

Vellata, C., Belli, S., Balsamo, F., Giordano, A., Colombo, R., & Maggioni, G. Effectiveness of Telerehabilitation on Motor Impairments, Non-motor Symptoms and Compliance in Patients With Parkinson's Disease: A Systematic Review. *Front Neurol.* 2021, 12:627999.

7 Technology-aided Programs to Support Leisure, Communication, and Daily Activities in People with Intellectual and Multiple Disabilities

Giulio E. Lancioni, Nirbhay N. Singh, Mark F. O'Reilly, Jeff Sigafoos, and Gloria Alberti

INTRODUCTION

People with mild to moderate or moderate to severe intellectual or multiple disabilities are frequently reported to have problems in managing independent access to leisure events, basic communication with partners not present in their immediate environment, and performance of functional daily activities (Badia et al., 2013; Davies et al., 2018; Desideri et al., 2021; Lancioni et al., 2018, 2020d; Lin et al., 2018). Their difficulties in accessing leisure events seem to be largely related to their inability to properly handle devices typically used for activating those events (e.g., music devices, computers, and tablets) (Chan et al., 2013; Lancioni et al., 2018), whereas their difficulties in managing basic communication exchanges with distant partners seem to be largely related to their inability to use telephone devices or comparable communication means (e.g., tablets and computers) instrumental for any of those exchanges (Darcy et al., 2017; Lancioni et al., 2016; Light et al., 2019). Finally, their problems in performing functional daily activities seem to be largely due to their inability to remember the steps involved in those activities and/or the sequence in which the steps are to be performed (Cannella-Malone & Schaefer, 2017; Desideri et al., 2021; Goo et al., 2019).

Given the negative implications of the aforementioned difficulties, the general consensus is that specific intervention programs are needed to help people find ways of alleviating such difficulties (Boot et al., 2018). With regard to those programs, the first point to consider is that they need to be conceived as strategies capable of

DOI: 10.1201/9781003272786-7

enabling people to reach independence rather than remaining passive or dependent on staff or caregivers' direct supervision. Intervention programs relying on direct supervision by staff or caregivers would, in fact, (a) perpetuate the persons' dependence rather than helping them reach new levels of initiative and self-determination, and (b) prove very expensive in terms of time and resources thus resulting barely affordable in many daily contexts (Lancioni et al., 2020a, 2020b; Wehmeyer, 2020; Wehmeyer et al., 2020). The second point is the obvious consequence of the first. In other words, the use of technology solutions may be a necessary and presumably practical approach to providing the level of support that people need to achieve their goals independent of staff and caregivers (Cullen et al., 2017; Desmond et al., 2018; Lancioni et al., 2018; Light et al., 2019).

In line with the aforementioned points, a variety of efforts has been made to develop technology-based programs that would foster people's independence. Generally, the programs were focused on fostering independence in one of the three main areas mentioned above, that is, (a) leisure (Lancioni et al., 2014, 2016, Stasolla et al., 2015, Wang et al., 2011), (b) communication (Kagohara et al., 2013, Ricci et al., 2017; van der Meer et al., 2012, 2017a, 2017b), or (c) performance of functional occupational/vocational activities (Desideri et al., 2021; Mechling et al., 2010; Savage & Taber-Doughty, 2017). The largely positive outcomes of the programs focusing on one specific area were viewed as a basis and an incentive for developing new programs that would target two of the areas or all three of them simultaneously. The reasoning was that programs targeting more areas could be more functional/satisfactory for the people involved and advantageous for the context running those programs. For example, providing people with the opportunity to freely shift from one type of engagement (e.g., leisure) to another (e.g., communication) could be relevant in increasing their range of opportunities and thus improve the quality of their engagement and their personal satisfaction. Setting up programs that would allow people to alternate periods of leisure and communication with periods of functional daily activity would ensure a combination of pleasant and functional/practical engagement that could (a) motivate people to remain positively occupied for relatively long periods of time and (b) have important benefits in terms of their overall level of independence, as well as in terms of staff and caregivers' time (Kazdin, 2012; Lancioni et al., 2020c, 2022).

The aim of this chapter is to provide an overview of selected studies that have assessed programs focusing on two or three areas simultaneously to explain the development of the programs, the technology solutions used to support them, and their outcomes in terms of participants' independent performance. Eight studies are summarized below. Six of the studies reported programs targeting two different areas, namely leisure and communication or leisure and functional daily activities. The other two studies reported programs targeting all three areas, that is, leisure, communication, and functional daily activities simultaneously. Following the presentation of the programs and their outcomes, a discussion section is provided that concentrates on three main issues: (a) the effectiveness of the programs and methodological considerations, (b) the accessibility and affordability of the programs, and (c) the implications of the programs for professionals working in daily contexts. With regard to the last issue, an effort was made to examine ethical and moral questions that may accompany the possible decisions of professionals to adopt those programs in daily contexts.

STUDIES TARGETING LEISURE AND COMMUNICATION OR LEISURE AND DAILY ACTIVITIES

The studies summarized in this section have been reported during the past few years and were aimed at helping adult participants with mild or moderate intellectual disabilities, often combined with sensory and/or motor impairments, to achieve independence in two of the three areas mentioned above, that is, in leisure and communication with distant partners or leisure and daily activities. The technology solutions adopted for the programs varied mainly on the basis of (a) the devices used to suit the characteristics of the participants (e.g., their different motor abilities and response ranges and their different sensory conditions) and (b) the way in which choice options were presented and the participants' response process was arranged (Federici & Scherer, 2017; Scherer, 2019).

Study I This study (Lancioni et al., 2017b) assessed a program based on the use of a laptop computer with screen and sound amplifier, a mobile communication modem, and a pressure microswitch with relative interface to help participants access various leisure options (e.g., songs, videos and slide shows) and manage communication exchanges with distant partners (i.e., via telephone calls or text messages). The participants were exposed to the program according to an ABAB design (with A and B representing baseline and intervention phases, respectively) and provided with 10-minute sessions across the different phases of the study. At the start of the sessions, the computer showed visual images of the three or four options available for the participant (e.g., images of singers/music, footballers, movies, and telephone), and illuminated and verbally presented each of them in succession. If the participant selected one of the options by activating the pressure microswitch during the program (or a mouse during the baseline), the computer presented a series of alternatives among which the participant could choose. For example, if the participant selected the singers/music option, the computer would present the titles of several songs and the participant could choose one of them via the microswitch (or the mouse during baseline). If the participant selected the telephone option, the computer would present the pictures and names of various relevant communication partners and the participant could, as mentioned above, choose one of them (i.e., the one to call or the one to contact via text message) through the microswitch or mouse response. Choosing a leisure event led the computer to present such an event (e.g., a song or video) for about two minutes. Choosing a partner to call led the computer to dial that partner's telephone number and start a telephone call. Choosing a partner to contact via text message led the computer to present various messages available for that partner and to send the message the participant chose immediately after the choice had occurred. The results showed that, during the baseline phases (i.e., when the mouse was available for use), the participants were unable to make any choices and thus could not access the leisure or communication options. During the intervention phases, they were successful in using the microswitch in connection with the computer system and were positively engaged in leisure and communication throughout the duration of the sessions.

In addition to the participants' leisure and communication engagement data, the study also reported a social validation of the program (carried out by means of a questionnaire presented to staff personnel) and an assessment of the participants' preferences (carried out by requesting them to choose between the program sessions and other presumably positive forms of engagement). The social validation indicated that staff had highly favorable ratings of the program. Moreover, the participants showed very strong preferences for the program sessions.

Study II This study (Lancioni et al., 2018) reported the use of (a) a smartphone with Android operating system, near-field communication, music and video player functions, and MacroDroid application, and (b) cards with radio-frequency identification tags. This technology was used for a program aimed at helping participants access different forms of preferred leisure events and make telephone calls with preferred communication partners. The program was introduced according to a non-concurrent multiple baseline design across participants, which involved 15-minute baseline and intervention sessions. During the baseline sessions, the research assistant (a) invited the participants to use a smartphone (which responded to specific verbal utterances and conventional touch inputs) and (b) pointed out the options accessible through the smartphone (e.g., listening to singers/songs and making telephone calls). During the intervention and post-intervention sessions, the participants had cards or miniature objects with radio-frequency identification tags. Those cards or miniature objects, which were familiar to the participants, represented leisure options as well as preferred communication partners that could be reached via a telephone call. The participants could access the corresponding leisure events or start calls with their preferred partners by holding the cards/objects onto the back of the smartphone. The smartphone would recognize the tags of those cards/objects through the MacroDroid application. Therefore, if the participant held a card/object representing a specific leisure event, the smartphone would deliver such an event. Similarly, if the participant held a card/object representing a specific communication partner, the smartphone would automatically set up a telephone call with that partner. During the baseline (without the support of the program), the participants were unable to request/access leisure events or start telephone calls independently. With the use of the program, they succeeded in doing both and spent most of their session time engaging in leisure and communication.

Study III. This study (Lancioni et al., 2020a) assessed a new technology-aided program to support independent leisure and communication engagement, which relied on the use of a tablet with Android operating system, a NANO SIM card, and the WhatsApp Messenger and MacroDroid applications. The participants were presented with (a) preferred leisure options and the possibility of communication with preferred partners via text messages, and, depending on the alternative they selected, (b) various steps they could take to access leisure events or send messages. Every program session lasted 10 min and started with the tablet sequentially scanning (illuminating for about 5 s) each of the two pictures representing the two choice areas (i.e., leisure and communication), and verbalizing each area while illuminating it. Some participants could select either area by approaching with their hand the proximity sensor of the tablet while that area was illuminated. Other participants (who could not exhibit such a motor response) were provided with a smartphone fixed to the headrest of their wheelchair for their selections. In practice, they could select the area they wanted by turning their head toward the smartphone (thus activating the smartphone's proximity sensor) when that area was illuminated. Following the selection of the area, the tablet presented the various alternatives available in that area (e.g., leisure alternatives, such as music and videos, or various communication partners to whom a message could be sent). Again, the participants were to use one of the aforementioned response strategies to select a leisure alternative or a communication partner. Based on this new selection, the tablet presented various options related to the leisure alternative selected (e.g., various singers or comedians) or listed various messages that could be sent to the partner selected. Selecting a leisure option (e.g., a singer) led the tablet to deliver an event related to it (e.g., a song) that the participants could enjoy. Selecting a message led the tablet to send it out. Incoming messages were automatically read by the tablet. Data showed that the participants were successful in using the technology-aided program and achieved high levels of independent leisure and communication engagement. A social validation of the program carried out by interviewing staff about the program's effective support of the participants (friendliness toward them) and applicability in daily contexts provided very positive results (i.e., staff rated the program highly on each of these aspects).

Study IV This study (Lancioni et al., 2020b) introduced a new technology solution that was seen as advantageous in terms of response requirement and/or choice process compared with the solutions used before. For example, the response required for making choices was relatively simple in relation to that required by using cards or miniature objects (see Study II). Moreover, the participants were no longer exposed to the successive scanning of the different options and events with the need to wait until the desired option/event was being scanned to make their choice (i.e., as occurred in Studies I and III). The technology used to achieve such an objective was the same as that reported for Study III (i.e., a tablet with Android operating system, a NANO SIM card, and the WhatsApp Messenger and MacroDroid applications). Yet, the program was set up in a different manner so that it would (a) present the participants with the choice elements in groups of three, and (b) allow the participants to choose the first, second, or third element of the group/sequence by touching or covering the tablet's proximity sensor once, twice, or three times.

Every intervention or post-intervention session (i.e., every session with the support of the technology-aided program) started with the tablet presenting verbally and visually three options (i.e., "music," "films," and "video calls"). The participant could choose the first, second or third option of the sequence by touching/covering the tablet's proximity sensor once, twice, or three times. If the participant chose music (i.e., covered the tablet's proximity sensor once), the tablet presented three music alternatives (i.e., three different singers). The participant could choose the first, second, or third singer using the same response strategy as before (i.e., touching/covering the tablet's sensor once, twice, or three times). If the participant refrained from choosing any of the singers, the tablet presented three additional singers. Once a singer had been chosen, the tablet presented three songs by that singer and the participant could choose one of them and listen to it. The same approach was also followed for the films option as well as the communication (video calls) option. If the participant chose video calls, the tablet presented three preferred communication partners that could be called. If the participant chose one of these partners, the tablet set up a video call with that partner through the WhatsApp Messenger. If the participant abstained from choosing any of the three partners, the tablet presented three additional partners among which the participant could choose. Data showed that the program was highly effective in helping the participants remain positively engaged in leisure and communication throughout the 10-min intervention and post-intervention sessions.

Study V In line with the studies summarized above, Study V (Lancioni et al., 2020c) focused on the evaluation of technology solutions for providing participants with the opportunity to access leisure events and manage communication exchanges with preferred partners not present in the immediate context. The technology system investigated in this study was designed to require fairly simple responses for its activation and thus to suit individuals with relatively serious developmental disabilities (which could also include blindness) and/or high levels of anxiety with fear of failure. The system allowed participants to make requests for leisure events or communication opportunities through simple hand-pressure responses. Those responses activated mini voice-recording devices, each of which contained a recorded voice message/request. The messages/requests recorded in the devices (and activated by the participants' hand-pressure responses) served to trigger the Google Assistant of a smartphone and hence to lead the smartphone to deliver what the messages/requests indicated. Eight devices were used during the intervention and post-intervention sessions. The requests of four of those devices concerned preferred leisure options, such as songs and comedy sketches. The requests of the other four devices concerned calls or text messages to specific (preferred) communication partners. Pictures or mini objects indicating those leisure options and partners were attached to the devices to make them easily discriminable to the participants. The system was set up in such a way that, if a communication partner was unreachable (could not respond to a call), the smartphone played a pre-recorded answer of that partner. Incoming messages would be automatically read by the smartphone. The results underlined the effectiveness of the program. In fact, the participants were not able to access leisure or communication without the full support of the system (i.e., during the baseline sessions). Following the introduction of the system, however, they were able to successfully engage in leisure and communication with consistency throughout the 10-min sessions.

The results of a social validation of the program conducted by interviewing staff personnel familiar with these education/rehabilitation issues were quite encouraging. Indeed, the staff interviewed rated the program (i.e., the technology system on which the program relied) as highly effective, enjoyable for the participants to use, suitable/functional for the participants' daily context, and worthy of being recommended.

Study VI The goal of this study (Lancioni et al., 2020d) was somewhat different from the goal of the studies summarized above. Indeed, this study was aimed at targeting leisure and functional daily activities. The technology involved a smartphone used in combination with special cards. The smartphone with Android operating system and near-field communication function was fitted with the MacroDroid application. The smartphone was also supplied with (a) audiovisual files concerning a variety of leisure events (e.g., comic sketches, travel adventures, and songs), as well as (b) pictorial images and verbal instructions concerning the activities to be carried out and the related steps. The cards were fitted with radio-frequency identification tags and represented the leisure options (i.e., as in Study II). The tags made the cards recognizable to the smartphone, which then delivered the corresponding leisure events during the intervention and post-intervention sessions (see Study II).

At the start of each of these sessions, the participant was sitting at a desk with the smartphone and eight cards representing eight different leisure options. The participant was expected to choose one of the cards and bring it in touch with the back of the smartphone. This led the smartphone to identify the card and play the corresponding leisure event. At the end of the event, the smartphone started to provide pictorial or verbal and pictorial instructions for one of the daily activities the participant was scheduled to carry out. The instructions were presented one at a time, automatically. Once the instructions for all activity steps had been presented, the smartphone invited the participant to go back to the cards and choose one of them. Choosing a card and bringing it to touch the back of the smartphone led the participant to access the leisure event related to that card. At the end of it, the smartphone presented a second activity (following the same process used for the first one). The session continued until the participant had accessed/enjoyed four different leisure events interspersed with three different activities. The results showed that, without the support of the program (i.e., during baseline sessions), the participants were unable to access leisure events and carry out daily activities independent of staff supervision. With the support of the program (i.e., during intervention sessions), the participants were able to manage independent choices of (and access to) leisure events and accurate performance of daily activities. In essence, the participants succeeded in alternating their engagement in leisure events with their engagement in daily activities and remained constructively occupied throughout sessions that lasted near to or more than 30 min.

STUDIES TARGETING LEISURE, COMMUNICATION, AND DAILY ACTIVITIES

The two studies summarized in this section were aimed at extending the impact of the technology-aided programs examined above. In essence, the new studies focused on all three main areas in which people with intellectual and multiple disabilities may need help (i.e., leisure, communication, and functional daily activities). An improvement in people's performance in those areas is considered to be critical for their general perspectives in terms of development, occupational achievement, and social adjustment. By extending the focus of the programs to all three areas, those studies were also able to generally expand the length of time the participants were remaining positively/constructively engaged in an independent manner.

Study VII The participants involved in this study (Lancioni et al., 2020e) were already familiar with a program enabling them to access leisure and communication. Indeed, the participants were able to use cards or miniature objects with tags in connection with a smartphone with Android operating system (as well as near-field communication function and the WhatsApp Messenger and the MacroDroid applications) to request leisure events, start telephone calls with preferred partners, or send those partners text messages. The purpose of the study was to extend the program in use with the introduction of activity periods similar to those available in Study VI. Sessions started with a period of about three minutes, during which the participants sat at a desk and could make leisure and communication choices. Once the 3-min period had elapsed and any leisure or communication event started within that period was completed, the smartphone presented the verbal instructions for the first activity. The instructions were presented one at a time, automatically, in support of the single steps of the activity so that the participants could perform those steps in the correct sequence. Following the last activity instruction, the smartphone informed the participants that they could start new leisure and communication choices. They were allowed a new 3-min period in which to make choices, which were as the ones described above. The end of this second leisure and communication choice period was followed by the instructions for a second activity. Sessions included four choice periods and three activities. The results showed that the participants were highly successful in integrating their leisure and communication engagement with the performance of daily activities during intervention and post-intervention sessions, which lasted close to or above 30 min.

A social validation assessment via staff interviews showed that a program such as the one carried out obtained significantly higher ratings than a program focusing on leisure and communication (i.e., such as those reported in Studies I–V). Higher ratings were provided on items concerning the range and variation of occupational opportunities, independent constructive occupation time, and contribution to the participants' rehabilitation process.

Study VIII This study (Lancioni et al., 2022) assessed the impact of a program relying on the use of a technology package comparable to that used in Study V, namely a package including (a) a smartphone with Android operating system, SIM card, Internet connection and Google account, and MacroDroid application, and (b) mini voice-recording devices. The participants were to use the voice-recording devices to request leisure events and activate telephone calls (i.e., as reported in Study V). The program was set up to alternate periods in which the participants could access leisure and carry out telephone calls (through the use of the aforementioned voice-recording devices, each of which had a recorded verbal message capable of activating the smartphone's Google Assistant) with periods in which daily activities were to be carried out. During the latter periods, the participants received the smartphone's instructions for the single activity steps. Data showed that the program was effective in enabling the participants to independently manage leisure, communication, and daily activities, and remain successfully engaged for sessions whose mean length varied between about 30 and close to 40 minutes.

EFFECTIVENESS OF THE PROGRAMS AND METHODOLOGICAL CONSIDERATIONS

With regard to the effectiveness of the programs, the data reported by the studies seem to be highly encouraging. The participants' steady leisure and communication engagement or the successful alternation of leisure and communication engagement with accurate performance of daily activities throughout the sessions in which the technology support was available may have a number of simple explanations. First, engaging in preferred leisure activities and communicating with preferred partners are most likely to represent highly motivating forms of engagement, and this may largely justify the participants' positive data reported by the different studies in those areas (Kazdin, 2012; Lancioni et al., 2017b, 2019). It may also be added that the possibility of switching between the two forms of engagement (i.e., between leisure and communication) could have played a positive role, by allowing the participants engagement variation. Indeed, the participants (a) were free to focus on what was more relevant for them at any particular time (King et al., 2014; Stasolla et al., 2015) and (b) could avoid any protracted use of the same form of engagement with possible declines in their interest levels. Second, the technology solutions employed in the various studies were adapted to the participants' characteristics and abilities, and thus were user-friendly and allowed the participants to achieve their goals in a fairly successful manner (i.e., avoiding failures and frustration). Third, the participants' successful performance of the activities scheduled was probably due to the fact that (a) the participants possessed the skills to carry out the steps of those activities, (b) the instructions provided were adequate to support the performance of those steps and ensure their correct sequencing, and (c) the participants were comfortable with the activity (instruction) situation and thus experienced a positive quality of engagement and avoided anxiety throughout their activity engagement time.

Although the results of the studies underline the effectiveness of the programs, some considerations may be needed about the research methodology used in the studies in order to determine the strength of those results. Regarding this point, it may be noted that the studies were carried out according to single-subject research designs (i.e., ABAB or non-concurrent multiple baselines across participants). While the ABAB is considered a strong design, the non-concurrent multiple baselines across participants may be viewed as a weaker design (i.e., a design providing less obvious evidence of the effects of the intervention; Krotochwill et al., 2013; Lobo et al., 2017). In spite of this apparent weakness, there are reasons to believe that the results obtained by the studies using such a design can still be considered valid (Lancioni et al., 2021). First, during the baseline, the participants showed zero or near-zero levels of performance in the areas targeted in contrast to the high levels they reached immediately after the introduction of the programs. Second, history or general environmental variables could not have accounted for those performance changes observed across participants over relatively short periods of time (Lancioni et al., 2021). In fact, those changes were not due to the acquisition of new language or motor skills but were related to the use of technology that substituted for those skills, which remained absent.

Another point, that should be taken into consideration in judging the results of the studies and their possible implications for daily contexts, is that all the studies summarized above were conducted by the same research group. This type of situation certainly limits the possibility of making general/conclusive statements about the evidence and could only be amended through the occurrence of replication studies carried out by different research groups (Kazdin, 2011; Locey, 2020; Travers et al., 2016). In reality, one such replicated study has just been carried out by a different research group, with adolescents emerging from a minimally conscious state (Stasolla et al., 2022). This study pursued independent leisure, communication, and occupation through a technology-aided program similar to those described in this chapter, and reported largely positive results, which strongly corroborate the evidence summarized above. In addition to receiving confirmation from this systematic replication study, the above evidence is also supported by the positive data from many simpler studies that successfully used technology-aided programs. Those studies were typically aimed at helping participants gain independence in one of the three areas addressed in this chapter, that is, leisure (Lancioni et al., 2016, Stasolla & De Pace, 2014 , Wang et al., 2011), communication (Kagohara et al. 2013, Lancioni et al., 2014; Ricci et al., 2017; van der Meer et al. 2012, 2017a, 2017b), and functional daily activities (Cullen et al., 2017; Lancioni et al., 2017a; Mechling et al., 2010).

ACCESSIBILITY AND AFFORDABILITY OF THE PROGRAMS

With regard to the issue of accessibility, the first question one can raise is whether the technology used for the programs can be directly obtained/purchased by different communities not connected with the research group that carried out the studies summarized above (Boot et al., 2018; de Witte et al., 2018; Federici & Scherer, 2017). Another question one can raise is whether the technology can be put into use in a straightforward manner within daily contexts. The answer to the first question is that all the technology employed in the studies can be directly purchased without any restrictions or need of connections with the research group responsible for its development and assessment. In fact, all components of the technology packages used within the programs are commercially available and thus directly accessible. The answer to the second question is not as simple as the answer to the first. In fact, while the technology components are commercially available, those components need to be arranged and programmed according to the objectives pursued for the participants. In essence, the technology packages are not ready-made (off-the-shelf) tools that one can directly use in the everyday context, but rather tools that one has to set up based on the scope of the intervention and the characteristics of the participants. This operation, albeit not too complex, may be beyond the skills of some staff and caregivers. Thus, staff and caregivers might need to ask for some assistance in the process.

With regard to the issue of affordability, the basic question concerns the cost of the packages reported by the studies summarized above. Given the differences among packages, the cost for their acquisition also varies. For example, packages including a smartphone in combination with cards or miniature objects fitted with radio-frequency identification tags are by far the cheapest and most easily affordable (Borg, 2019; Lancioni et al., 2020d). In fact, a smartphone adequate for the task can be acquired for probably about US$200, while the MicroDroid application, tags, cards, and miniature objects have negligible costs with hardly any impact on affordability. The cost of the packages will obviously increase when a tablet or a combination of a tablet and a smartphone are used within the same program (Study III).

PERSON-CENTERED IMPLICATIONS OF THE PROGRAMS (SOCIAL AND ETHICAL/MORAL CONSIDERATIONS ON PROGRAM ADOPTION)

The positive results of the studies summarized and the aforementioned discussion on the issues of program effectiveness, research methodology, program accessibility, and program affordability would seem to make a case in favor of the adoption and use of those programs in daily contexts. Obviously, those contexts may need to have extra evidence (e.g., may want to see or be involved in replication studies conducted by different research groups) before committing themselves to implementing such programs as regular rehabilitation tools (Campbell et al., 2014; Campos et al., 2019; Locey, 2020). That commitment may depend on the outcome of the aforementioned replication studies and on the answers to a number of questions that may constitute a social and ethical/moral framework within which a decision on those programs eventually needs to be taken. At least five immediate questions could be envisaged as part of such a framework. First, do staff, caregivers or other service providers and stakeholders consider those programs effective and relevant for the participants' rehabilitation progress and the achievement of a better life condition (i.e., for enhancing the participants' well-being)? Second, do participants show interest and enjoyment during those programs? Third, could the participants reach the same objectives without the use of those programs? Fourth, what would be the most likely alternatives to the programs within daily contexts? Fifth, are there social and ethical/moral duties that service providers (e.g., rehabilitation and research personnel) should consider as binding for their decisions following new research (replication studies) and clarifying answers to the aforementioned questions?

While waiting for new/independent replication studies, a number of preliminary, plausible answers might be provided in relation to each of the questions listed above. Regarding the first question (i.e., staff, caregivers, or other service providers and stakeholders' view of the programs), some evidence is already available. In four of the studies summarized above (i.e., Studies I, III, V, and VII) and in the study by Stasolla et al. (2022), social validations of the programs were reported. In those social validations, staff and other relevant rehabilitation and care personnel were interviewed about the impact of the programs in supporting participants' performance and progress as well as about the applicability of the programs. The personnel's ratings relative to all those aspects were quite favorable, suggesting that they had a positive view of the programs' potential and usability.

Regarding the second question (i.e., participants' interest for and enjoyment of the programs), the literature has provided some specific evidence/data as well as anecdotal reports. Indeed, Lancioni et al. (2017b) carried out preference checks in which the participants could choose whether to be involved in a program session or in another form of activity considered to be positive/pleasant for them. Their data showed that the participants preferred to be involved in the program sessions almost consistently. Similarly, Lancioni et al. (2020b, c) and Stasolla et al. (2022) provided anecdotal reports about the participants showing signs of pleasure and satisfaction during the program sessions. One more point that might support the notion of participants' interest for and enjoyment of the program sessions is the high and consistent level of engagement that they displayed throughout the typically extended program assessment periods reported by the studies summarized in this chapter, as well as by other studies (e.g., Lancioni et al., 2019, 2020f; Stasolla et al., 2022).

Regarding the third question (i.e., participants' chances of pursuing the goals of independent leisure, communication, and daily activities without the use of those programs), no direct answer exists. Nonetheless, three points may be put forward which suggest that such an achievement could hardly be obtained independent of the programs. First, all the participants involved in the studies summarized above were adults and had been attending rehabilitation and care facilities without reaching independent leisure and communication or leisure, communication, and daily activities. Second, given the participants' general conditions, it is most unlikely that they would (or could) have developed the motor and memory skills necessary to achieve the results reported through typical rehabilitation practices or conventional technology (Lancioni et al., 2019, 2021). Third, the programs allowed the participants to bypass their difficulties/weaknesses by bridging the gap between their real skills and the skill level requested of them to achieve independence in the targeted areas.

Regarding the fourth question (i.e., possible/likely alternatives to the programs that one could use within daily contexts), no specific data or answers are available. The most reasonable assumption (also in light of the history of the participants included in the studies summarized above) is that the participants would be provided with some form of assistance by staff or caregivers so as to allow them to reach, at least in part, the objectives achievable through the technology-aided programs. Obviously, the use of staff and caregiver support would have two disadvantages: it would perpetrate a level of participants' dependence, and would have clear costs in terms of staff and caregivers' time and effort (Stasolla et al., 2022; Wehmeyer, 2020).

Regarding the fifth question (i.e., the social and ethical/moral duties of staff, service providers, and rehabilitation and research personnel), the best response one can give is that the decisions and practices of all these professionals should be in line (a) with the data currently available, the results of new, independent replication studies as well as the answers provided to the questions examined above, and (b) with their commitment to offer the participants the most effective and satisfactory form of support possible (Adams & Boyd, 2010; Man & Kangas, 2020; McMcDonald et al., 2015). Accepting that the evidence and answers provided above suggest that the programs can have a good chance of contributing positively to improve the participants' situation would have social and ethical/moral implications for professionals in the area. That is, those professionals would be required to consider the notion (and probably feel the social and ethical/moral duty) of working toward adapting and accommodating the programs for use within their premises.

CONCLUSIONS

This chapter summarizes some recent studies evaluating technology-aided programs to support participants with intellectual and other disabilities to independently manage leisure and communication or combinations of leisure, communication, and daily activities. The encouraging results reported by the studies and the relative accessibility and affordability of the technology solutions adopted to support the programs have prompted a discussion concerning the strength of the evidence (and the possibility to generalize such evidence to daily contexts). In debating this point, considerations were made about (a) methodological issues that may raise some caution against drawing direct, strong conclusions from the evidence provided by the studies summarized, and (b) the large amount of data on technology-aided programs that justifies an attitude quite favorable toward such an approach and hopeful as to its beneficial impact for a growing number of persons in need. The final points analyzed were concerned with person-centered social and ethical considerations that staff, caregivers, and other service providers can hardly avoid. Those considerations may provide a new impetus toward preparing daily contexts to adopt and sustain technology-aided programs aimed at promoting participants' independence in leisure, communication, and daily activities.

REFERENCES

Adams, Z. W., & Boyd, S. E. (2010). Ethical challenges in the treatment of individuals with intellectual disabilities. *Ethics & Behavior, 20*(6), 407–418.

Badia, M., Orgaz, M. B., Verdugo, M. A., & Ullán, A. M. (2013). Patterns and determinants of leisure participation of youth and adults with developmental disabilities. *Journal of Intellectual Disability Research, 57*(4), 319–332.

Boot, F. H., Owuor, J., Dinsmore, J., & MacLachlan, M. (2018). Access to assistive technology for people with intellectual disabilities: A systematic review to identify barriers and facilitators. *Journal of Intellectual Disability Research, 62*(10), 900–921.

Borg, J. (2019). Commentary on selection of assistive technology in a context with limited resources. *Disability and Rehabilitation: Assistive Technology, 14*(8), 753–754.

Campbell, M., Robertson, A., & Jahoda, A. (2014). Psychological therapies for people with intellectual disabilities: Comments on a matrix of evidence for interventions in challenging behaviour. *Journal of Intellectual Disability Research, 58*(2), 172–188.

Campos, R. C., Holden, R. R., & Lambert, C. E. (2019). Avoidance f psychological pain and suicidal ideation in community samples: Replication across two countries and two languages. *Journal of Clinical Psychology, 75*(12), 2160–2168.

Cannella-Malone, H. I., & Schaefer, J. M. (2017). A review of research on teaching people with significant disabilities vocational skills. *Career Development and Transition for Exceptional Individuals, 40*(2), 67–78.

Chan, J. M., Lambdin, L., Van Laarhoven, T., & Johnson, J. W. (2013). Teaching leisure skills to an adult with developmental disabilities using a video prompting intervention package. *Education and Training in Autism and Developmental Disabilities*, *48*(3), 412–420.

Cullen, J. M., Alber-Morgan, S. R., Simmons-Reed, E. A. and Izzo, M. V. 2017. Effects of self-directed video prompting using iPads on the vocational task completion of young adults with intellectual and developmental disabilities. *Journal of Vocational Rehabilitation*, *46*, 361–375.

Darcy, S., Green, J., & Maxwell, H. (2017). I've got a mobile phone too! Hard and soft assistive technology customization and supportive call centres for people with disability. *Disability and Rehabilitation Assistive Technology*, *12*(4), 341–351.

Davies, D. K., Stock, S. E., Herold, R. G., & Wehmeyer, M. L. (2018). GeoTalk: A GPS-enabled portable speech output device for people with intellectual disability. *Advances in Neurodevelopmental Disorders*, *2*(3), 253–261.

Desideri, L., Lancioni, G., Malavasi, M., Gherardini, A., & Cesario, L. (2021). Step instruction technology to help people with intellectual and other disabilities perform multistep tasks: A literature review. *Journal of Developmental and Physical Disabilities*, *33*(6), 857–886.

Desmond, D., Layton, N., Bentley, J., Boot, F. H., Borg, J., Dhungana, B. M., Gallagher, P., Gitlow, L., Gowran, R. J., Groce, N., Mavrou, K., Mackeogh, T., McDonald, R., Pettersson, C., & Scherer, M. J. (2018). Assistive technology and people: A position paper from the first global research, innovation and education on assistive technology (GREAT) summit. *Disability and Rehabilitation: Assistive Technology*, *13*(5), 437–444.

de Witte, L., Steel, E., Gupta, S., Delgado Ramos, V., & Roentgen, U. (2018). Assistive technology provision: Towards an international framework for assuring availability and accessibility of affordable high-quality assistive technology. *Disability and Rehabilitation: Assistive Technology*, *13*(5), 467–472.

Federici, S., & Scherer, M. J. (Eds.) (2017). *Assistive technology assessment handbook* (2nd ed.). London: CRC Press.

Goo, M., Maurer, A. L., & Wehmeyer, M. L. (2019). Systematic review of using portable smart devices to teach functional skills to students with intellectual disability. *Education and Training in Autism and Developmental Disabilities*, *54*(1), 57–68.

Kagohara, D. M., van der Meer, L., Ramdoss, S., O'Reilly, M. F., Lancioni, G. E., Davis, T. N., et al. (2013). Using iPods and iPads in teaching programs for individuals with developmental disabilities: A systematic review. *Research in Developmental Disabilities*, *34*(1), 147–156.

Kazdin, A. E. (2011). *Single-case research designs: Methods for clinical and applied settings* (2th ed.). New York: Oxford University Press.

Kazdin, A. E. (2012). *Behavior modification in applied settings* (7th ed.). New York: Waveland Press.

King, G., Gibson, B. E., Mistry, B., Pinto, M., Goh, F., Teachman, G., & Thomposn, L. (2014). An integrated methods study of the experiences of youth with severe disabilities in leisure activity settings: The importance of belonging, fun, and control and choice. *Disability and Rehabilitation*, *36*(19), 1626–1635.

Kratochwill, T. R., Hitchcock, J. H., Horner, R. H., Levin, J. R., Odom, S. L., Rindskopf, D. M., & Shadish, W. R. (2013). Single case intervention research design standards. *Remedial and Special Education*, *34*(1), 26–38.

Lancioni, G. E., Singh, N., O'Reilly, M., Sigafoos, J., Boccasini, A., La Martire, M. L., & Lang, R. (2014). Case studies of technology for adults with multiple disabilities to make telephone calls independently. *Perceptual and Motor Skills*, *119*(1), 320–331.

Lancioni, G. E., O'Reilly, M., Singh, N., Sigafoos, J., Boccasini, A., La Martire, M. L., Perilli, V., & Spagnuolo, C. (2016). Technology to support positive occupational engagement and communication in persons with multiple disabilities. *International Journal on Disabilities and Human Development*, *15*(1), 111–116.

Lancioni, G. E., Singh, N. N., O'Reilly,M. F., Sigafoos, J., Alberti, G., Zimbaro, C., & Chiariello, V. (2017a). Using smartphones to help people with intellectual and sensory disabilities perform daily activities. *Frontiers in Public Health*, *5*, 282. https://doi.org/10.3389/fpubh.2017.00282.

Lancioni, G. E., Singh, N. N., O'Reilly, M. F., Sigafoos, J., Boccasini, A., Perilli, V., & Spagnuolo, C. (2017b). Persons with multiple disabilities manage positive leisure and communication engagement through a technology-aided progarm. *International Journal of Developmental Disabilities*, *63*(3), 148–157.

Lancioni, G. E., Singh, N. N., O'Reilly, M. F., Sigafoos, J., Alberti, G., Perilli, V., Chiariello, V., & Buono, S. (2018). An upgraded smartphone-based program for leisure and communication of people with intellectual and other disabilities. *Frontiers in Public Health*, *6*, 234. https://doi.org/10.3389/fpubh.2018.00234.

Lancioni, G. E., Singh, N. N., O'Reilly, M. F., Alberti, G., Chiariello, V., Campanella, C., Grillo, G., & Tagliente, V. (2019). A program based on common technology to support communication exchanges and leisure in people with intellectual and other disabilities. *Behavior Modification*, 43(6), 879–897.

Lancioni, G. E., Singh, N. N., O'Reilly, M. F., Sigafoos, J., Alberti, G., Perilli, V., Chiariello, V., Grillo, G., & Turi, C. (2020a). Case series of technology-aided interventions to support leisure and communication in extensive disabilities. *International Journal of Developmental Disabilities*, *66*(3), 180–189.

Lancioni, G. E., Singh, N. N., O'Reilly, M. F., Sigafoos, J., Grillo, G., Desideri, L., Alberti, G., & Campodonico, F. (2020b). A new tablet-based program to support leisure and video calls in people with intellectual and motor disabilities. *Technology and Disability*, *32*(2), 111–121.

Lancioni, G. E., Singh, N. N., O'Reilly, M. F., Sigafoos, J., Alberti, G., Chiariello, V., & Desideri, L. (2020c). People with intellectual and visual disabilities access basic leisure and communication using a smartphone's google assistant and voice recording devices. *Disability and Rehabilitation: Assistive Technology*. https://doi.org/10.1080/17483107.2020.1836047

Lancioni, G. E., Singh, N. N., O'Reilly, M. F., Sigafoos, J., Alberti, G., Chiariello, V., & Carrella, L. (2020d). Everyday technology to support leisure and daily activities in people with intellectual and other disabilities. *Developmental Neurorehabilitation*, *23*(7), 431–438.

Lancioni, G. E., Singh, N. N., O'Reilly, M. F., Sigafoos, J., Alberti, G., Chiariello, V., & Buono, S. (2020e). Extended smartphone-aided program to sustain daily activities, communication and leisure in individuals with intellectual and sensory-motor disabilities. *Research in Developmental Disabilities*, *105*, 103722. https://doi.org/10.1016/j.ridd.2020.103722

Lancioni, G. E., Singh, N. N., O'Reilly, M. F., Sigafoos, J., Alberti, G., Perilli, V., Chiariello, V., Grillo, G., & Turi, C. (2020f). A tablet-based program to enable people with intellectual and other disabilities to access leisure activities and video calls. *Disability and Rehabilitation: Assistive Technology*, *15*(1), 14–20.

Lancioni, G. E., Desideri, L., Singh, N. N., Sigafoos, J., & O'Reilly, M. F. (2021). A commentary on standards for single-case experimental studies. *International Journal of Developmental Disabilities*. https://doi.org/10.1080/20473869.2020.1870420

Lancioni, G. E., Singh, N. N., O'Reilly, M. F., Sigafoos, J., Alberti, G., Campodonico, F., & Desideri, L. (2022). A smartphonebased program enabling people with intellectual and other disabilities to access leisure, communication, and functional activities. *Universal Access in the Information Society*. https://doi.org/10.1007/s10209-021-00858-4

Light, J., McNaughton, D., & Caron, J. (2019). New and emerging AAC technology supports for children with complex communication needs and their communication partners: State of the science and future research directions. *Augmentative and Alternative Communication*, *35*(1), 26–41.

Lin, M. L., Chiang, M. S., Shih, C. H., & Li, M. F. (2018). Improving the occupational skills of students with intellectual disability by applying video prompting combined with dance pads. *Journal of Applied Research in Intellectual Disabilities*, *31*(1), 114–119.

Lobo, M. A., Moeyaert, M., Baraldi Cunha, A. and Babik, I. 2017. Single-case design, analysis, and quality assessment for intervention research. *Journal of Neurologic Physical Therapy*, *41*(3), 187–197.

Locey, M. L. (2020). The evolution of behavior analysis: Toward a replication crisis? *Perspectives on Behavior Science*, *43*(4), 655–675.

Man, J., & Kangas, M. (2020), Best practice principles when working with individuals with intellectual disability and comorbid mental health concerns. *Qualitative Health Research*, *30*(4), 560–571.

McDonald, K. E., Schwartz, N. M., Gibbons, C. M., & Olick, R. S. (2015). "You can't be cold and scientific": Community views on ethical issues in intellectual disability research. *Journal of Empirical Research on Human Research Ethics*, *10*(2), 196–208.

Mechling, L. C., Gast, D. L., & Seid, N. H. (2010). Evaluation of a personal digital assistant as a self-prompting device for increasing multi-step task completion by students with moderate intellectual disabilities. *Education and Training in Autism and Developmental Disabilities*, *45*(3), 422–439.

Ricci, C., Miglino, O., Alberti, G., Perilli, V., Lancioni, G.E. (2017). Speech generating technology to support request responses of persons with intellectual and multiple disabilities. *International Journal of Developmental Disabilities*, *63*(4), 238–245.

Savage, M. N., & Taber-Doughty, T. (2017). Self-operated auditory prompting systems for individuals with intellectual disability: A meta-analysis of single-subject research. *Journal of Intellectual & Developmental Disability*, *42*(3), 249–258.

Scherer, M. J. (2019). Assistive technology selection to outcome assessment: The benefit of having a service delivery protocol. *Disability and Rehabilitation: Assistive Technology*, *14*(8), 762–763.

Stasolla, F., & De Pace, C. (2014). Assistive technology to promote leisure and constructive engagement by two boys emerged from a minimal conscious state. *Neurorehabilitation*, *35*(2), 253–259.

Stasolla, F., Perilli, V., Di Leone, A., Damiani, R., Albano, V., Stella, A., & Damato, C. (2015). Technological aids to support choice strategies by three girls with rett-syndrome. *Research in Developmental Disabilities*, *36*, 36–44.

Stasolla, F., Caffò, A. O., Bottiroli, S., & Ciarmoli, D. (2022). An assistive technology program for enabling five adolescents emerging from a minimally conscious state to engage in communication, occupation, and leisure opportunities. *Developmental Neurorehabilitation*, *25*(3), 193–204.

Travers, J. C., Cook, B. G., Therrien, W. J., & Coyne, M. D. (2016). Replication research and special education. *Remedial and Special Education*, *37*(4), 195–204.

van der Meer, L., Kagohara, D., Achmadi, D., O'Reilly, M. F., Lancioni, G. E., Sutherland, D., & Sigafoos, J. (2012). Speech-generating devices versus manual signing for children with developmental disabilities. *Research in Developmental Disabilities*, *33*(5), 1658–1669.

van der Meer, L., Matthews, T., Ogilvie, E., Berry, A., Waddington, H., Balandin, S., O'Reilly, M. F., Lancioni, G., Sigafoos, J. (2017a). Training direct-care staff to provide communication intervention to adults with intellectual disability: A systematic review. *American Journal of Speech-Language Pathology*, *26*(4), 1279–1295.

van der Meer, L., Waddington, H., Sigafoos, J., Balandin, S., Bravo, A., Ogilvie, E., Matthews, T., & Sawchak, A. (2017b). Training direct-care staff to implement an iPad®-based communication intervention with adults with developmental disability. *International Journal of Developmental Disabilities, 63*(4), 246–255.

Wang, S. H., Chiang, C. S., Su, C. Y., & Wang, C. C. (2011). Effectiveness of virtual reality using Wii gaming technology in children with down syndrome. *Research in Developmental Disabilities, 32*(1), 312–321.

Wehmeyer, M. L. (2020). The importance of self-determination to the quality of life of people with intellectual disability: A perspective. *International Journal of Environmental Research and Public Health, 17*(19), 7121. https://doi.org/10.3390/ijerph17197121

Wehmeyer, M. L., Davies, D. K., Stock, S. E., & Tanis, S. (2020). Applied cognitive technologies to support the autonomy of people with intellectual and developmental disabilities. *Advances in Neurodevelopmental Disorders, 4*(4), 389–399.

8 The Role of Healthcare and Social Care Professionals in Supporting Access to and Engagement with Healthcare Technologies

Helen Hawley-Hague, Ellen Martinez, Norina Gasteiger, Claire Ford, and Emma Stanmore

INTRODUCTION

The World Health Organization (WHO) has a strategic vision for digital health to "be supportive of equitable and universal access to quality health services" through its Global Strategy on Digital Health (WHO, 2021). The NHS Long-Term Plan in the United Kingdom (UK) recommends "digital enablement" and that health professionals should have appropriate tools to support patients (NHS England, 2019). Health professionals are identified as important gatekeepers to successful engagement with health technology across the world, providing access, confidence, expertise, and support (Odendaal et al., 2020). Health professionals can play a mediating effect in relation to patients' technology self-efficacy (i.e., the belief that they can engage with and/or carry out an activity) (Graffigna et al., 2016), such as through supporting patients to become more digitally literate or signposting to evidence-based technologies.

There are several theoretical models that could be used to assess an individual's approach toward adopting new technologies. Health professionals do not work alone and their acceptance of technology and their influence on their patients or clients can be understood in the context of frameworks of technology acceptance. Greenhalgh et al. (2014) have developed a theoretical model of health professional resistance to new healthcare technology. This includes four key factors: resistance to the **policy** reflected in the technology (e.g., a policy of shifting the work of disease management from professional to patient); resistance to the **socio-material constraints**

DOI: 10.1201/9781003272786-8

(e.g., clunkiness) of the new technology; resistance to **compromised professional practice** (e.g., less scope for exercising judgment); and resistance to **compromised professional relationships** (e.g., a perception that a remote interaction is less professional than a face-to-face one). More recently, they have developed a broader framework that could be applied to better understand organizational technology acceptance and diffusion, known as the Nonadoption, Abandonment, Scale-Up, Spread, and Sustainability (NASSS) framework. This considers seven domains that can be applied to the adoption of technology, which includes adopters (including health professionals' role and identity) and an organizational domain (including capacity to innovate, readiness, and extent of change) as well as a condition domain which includes healthcare conditions, aging and deprivation as examples (Greenhalgh et al., 2017). The Technology Acceptance Model (TAM) also provides a framework to assess individual health and social care professionals' attitudes towards technology. It focuses on whether a technology is perceived by the user as useful, and whether it is easy to use (Davis, 1989).

Health and social care professionals' roles, attitudes, knowledge, and expectations, as well as the support and readiness to implement change from their organizations and individual conditions of the patients they are working with, are intertwined (Greenhalgh et al., 2017).

In this chapter, we discuss the use of a range of different technologies in healthcare including teleconferencing, mHealth, exergames, virtual reality, and robotics. Specifically, we focus on the opportunities, shortfalls, and roles of health and social care professionals in facilitating the implementation and uptake of the technologies within healthcare, particularly when working with older and frail populations. In doing so, we identify key barriers and facilitators to technology adoption faced by health and social care professionals and patients, using delivery with older adults or those with long-term conditions (condition domain) as primary examples.

TELECONFERENCING AND MHEALTH

mHealth (mobile health) has been found to support health professionals in delivering healthcare and preventative health (Odendaal et al., 2020; Rowland et al., 2020). mHealth is defined by the World Health Organization as the "use of mobile and wireless technologies to support the achievement of health objectives" (World Health Organization, 2012).

Smartphones, for example, are portable and can provide support and feedback at any time with the potential to improve motivation and compliance with healthcare interventions or self-management; they also do not tend to be switched off in the same way that other devices are (del Rosario et al., 2015; King et al., 2013; Mellone et al., 2012). mHealth can be used to detect vital signs, motivate activity, self-manage conditions, or deliver therapy either through apps or teleconferencing platforms (Rowland et al., 2020). Patients have highlighted that remote teleconferencing or app use should supplement rather than replace contact with health and social care professionals (Abramsky et al., 2018; Levinger et al., 2017; Muller et al., 2015; Vo et al., 2019). For the delivery of physiotherapy, mHealth has been found to enhance the therapeutic relationship by creating an interactive and on-going connection between patients and their physiotherapists (Abramsky et al., 2018; Cottrell et al., 2018; Levinger et al., 2017; Muller et al., 2015).

HEALTH PROFESSIONALS' EXPERIENCES AND ATTITUDES TOWARD mHEALTH

Knowledge and limited digital literacy at the individual health and social care professional levels have been identified as being important in the engagement of such personnel with mHealth (Odendaal et al., 2020, Zakerabasali et al., 2021), with the right support and training raised as being important on an organizational and individual level (Gagnon et al., 2016; Hawley-Hague et al., 2022, 2023).

The impact on capacity, workload, and working conditions is also an important mediator in decisions around whether to support mHealth interventions, particularly if it is in addition to in-person care. For example, in a recent study, physiotherapists highlighted the potential impact on their workload of using remote physiotherapy if they were expected to provide it in addition to in-person care (Hawley-Hague et al., 2022, 2023). In our evaluation of remote physiotherapy during the COVID-19 pandemic, both survey and interview data highlighted that delivering remote physiotherapy was very intense, requiring great concentration and regular breaks. Some staff likened it to working in a 'call center' (Hawley-Hague et al., 2022, 2023).

Health professionals have expressed concerns about patient safety if they are not physically present to assist a patient (Ramage et al., 2021), or of "missing something" if they do not have physical contact (Hawley-Hague et al., 2022; Malliaras et al., 2021; Ramage et al., 2021). This has been identified as particularly challenging when working with people with mobility and balance problems, or who are at risk of falls (Hawley-Hague et al., 2022; Hawley-Hague et al., 2021), and as a challenge to patient-centered delivery. Risks to patient privacy and confidentiality are another barrier raised by health professionals in relation to delivery (Cottrell et al., 2019; Gagnon et al., 2016).

The challenge to professional identity is identified as a barrier to health and social care professionals' engagement with mHealth, teleconferencing, and patient-facing technologies (Gagnon et al., 2016; Hawley-Hague et al., 2022). For example, in-person care and physical touch have been identified as essential parts of nursing and physiotherapy and health professionals have voiced concerns that virtual consultations would lead to loss of **"hands-on skills,"** leading to a decrease in specialist skills and poorer patient care (Hawley-Hague et al., 2022, 2023).

Large numbers of health professionals can also see the benefits of mHealth and perceive remote assessment of need or delivery of healthcare as an opportunity to be innovative and **"think outside the box,"** gaining new skills. Some health professionals, for example, have seen the COVID-19 pandemic as a catalyst for positive change (Hawley-Hague et al., 2022, 2023).

Confidence in using the technology has been found to be important to both health professionals and patients. Healthcare professionals are more likely to adopt technological interventions if they perceive the benefit to themselves and patients (MacNeill et al., 2014). Healthcare professionals have seen their confidence as important in encouraging patient confidence in the use of the technology (Hawley-Hague et al., 2020). There has also been a perception that the more both health professionals and patients use mHealth, the more confident they become (Hawley-Hague et al., 2022; Vaportzis et al., 2017). Pro-active and capable leadership has been found to positively influence the experience of the introduction of mHealth and remote aspects of healthcare, particularly in the support of within-team confidence and self-efficacy, which are also linked to opportunities and the availability of training.

Health professionals have stated that remote consultations have been particularly useful for initial triage and screening appointments to direct patients to the most appropriate type of care and complete subjective assessments (Hawley-Hague et al., 2022). They have also been found to reduce pressure on acute services and primary care (Bunn et al., 2004). Telephone assessments has quite an advanced existing literature related to the opportunity they can provide for more focused and efficient healthcare (Salisbury et al., 2013).

Video conferencing has not only been found to have advantages for healthcare delivery directly in interactions with patients. It has also been found to be a highly cost-effective method of facilitating multidisciplinary team meetings within teams of health professionals and provides easier access to external expertise centers (Dulai et al., 2020). In our UK-wide evaluation, it was found to be very useful in the support of rehabilitation teams across a range of disciplines and gave multidisciplinary teams better access to support from consultants, as well as involving less travel and fewer parking challenges (Aghdam et al., 2019; Hawley-Hague et al., 2022).

EXERGAMES

Exergames are increasingly being recognized as having the potential to increase engagement using targeted exercises, or to increase general physical activity in older populations. Exergames are a type of gamified telerehabilitation that combines animated gameplay (similar to videogames) with physical exercise. Rehabilitation outcomes are dependent on the adherence of the patient to the prescribed therapy (Piech & Czernicki, 2021).

Exergame systems (when well designed with the involvement of users and based on best evidence) have the potential to increase adherence through motivational design elements (competition, rewards, feedback, and immersion), and can provide continuous monitoring of use and interaction with the user to ensure that correct movements and adjustments are safely made. Most exergames utilize camera sensors, such as the Microsoft Kinect or the Orbbec 3D motion tracking devices, that do not require handheld controls. Individual exercise programs can be tailored using a choice of games and lower or upper limb therapy-based exercises. The exergame system can track the users' performances and measure parameters including: the frequency and duration of exergames played; progress statistics including range of motion, improvement scores, distance, speed, acceleration; and overall activity during the games (Stanmore et al., 2019).

The motivational aspects of exergames are due to the user being focused on the competitive goals or becoming immersed in the gameplay rather than focusing on

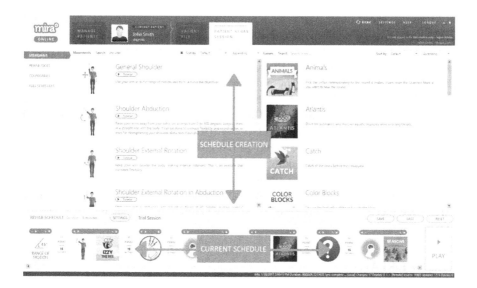

FIGURE 8.1 Example of an Exergame schedule with a personalized program of exercises are matched to specific games where gameplay facilitates therapeutic movements (reproduced with permission, MIRA Rehab Ltd).

the repetitive element of the exercise (Yen & Chiu, 2021). Some exergame systems also support remote monitoring, whereby the users' exercise program and progress in the home setting can be viewed and adjusted by the supervising clinician in the clinic or in-patient environment (see Figure 8.1).

HEALTH PROFESSIONALS' EXPERIENCES AND ATTITUDES TOWARD EXERGAMES

The implementation of exergames may present a number of challenges to health and social care professionals. Some of these barriers apply to health technologies more widely as discussed earlier (digital literacy, access to hardware, data security, and interaction with health records) but others are specifically related to exergames. These include issues such as the portability of the device, who will maintain and pay for the software licence(s), and the complexity of the individual systems and sensitivity of the sensors, as well as the appropriateness and acceptability of the animated games associated with the exercises (Stanmore et al., 2019).

A recent review of exergames has highlighted their ability to assist health and social care professionals in making the best use of resources, enabling older adults to self-manage as optimally as possible, and their potential to increase engagement with evidence-based exercises to support older adults with disabilities, functional decline, motor and cognitive impairments for rehabilitation, pre-rehabilitation, falls, and frailty prevention services (Meulenberg et al., 2022).

In research with older adults, exergames have been demonstrated to be effective in reducing falls and improving balance, pain and confidence after 12 weeks use in assisted-living facilities in the UK (Stanmore et al., 2019). In healthy older adults, exergames have been shown to improve balance and be as effective, if not more effective, than traditional therapy programs (van Diest et al., 2013; Wuset et al., 2014). Health and social care workers have also noted the motivational and enjoyable aspects of exergames (Nawaz et al., 2016), as well as their use for rehabilitation and recovery (Kiziltas & Celikcan, 2018), post-stroke (Koh et al., 2020), neurological conditions (Rosly et al., 2017), cognitive impairment (Stanmore et al., 2017), and frailty (Zheng et al., 2019). In addition, an area of particular interest in the neuroscience research community is the dual-task training components of exergames that may target brain plasticity and cognitive functioning as well as physical mobility (Meulenberg et al., 2022), enabling health professionals to provide multiple benefits to patients.

VIRTUAL REALITY

Virtual Reality (VR) is another emerging technology being utilized in the health and social care sector. VR is a computer-generated, three-dimensional, multi-sensory, immersive environment or image which can be interacted with by a user through specialist equipment, such as head-mounted displays, handsets, or haptics (Cipresso et al., 2018; Skarbez et al., 2018). VR is being developed and implemented in numerous clinical settings and utilized for different patient groups for managing and assisting with numerous conditions through improving the patient experience (Bekelis et al., 2017) and patient outcomes (Winter et al., 2021). Examples include managing anxiety disorders (Chesham et al., 2018), supporting physical rehabilitation (Mirelman et al., 2016; de Rooj et al., 2016), pain management (Ahmadpour et al., 2019), and distractions while delivering care interventions (Malloy & Milling, 2010). All of these examples require the facilitation of a healthcare provider to assist the user.

VR is used to assist older adults with neurological conditions, including Parkinson's disease and stroke, in gait rehabilitation, as well as for those with multiple sclerosis (Mirelman et al., 2016; de Rooj et al., 2016; Winter et al., 2021). VR-based treadmill training assists in improving gait, stride length, stride speed, and walking endurance (Mirelman et al., 2016; de Rooi et al., 2016; Winter et al., 2021). This can improve functional independence, reducing the risk of falls, potential future injuries, and delaying progression of degenerative conditions (Meldrum et al, 2012; Mendes et al., 2012). VR is also used to assist patients with post-traumatic stress disorder (PTSD) through controlled management of exposure therapy (Kothgassner et al., 2019). Through repeated confrontation with the trigger, being able to control the stimulus and being exposed to the repetition of using the VR headset is beneficial in exposure therapies. Similarly, VR is also being used in exposure therapies as a treatment for patients with specific phobias (Freitas et al., 2021) and addiction disorders (Mazza et al., 2021). VR can give the user a sense of presence, which improves user outcomes (Botella et al., 2017).

Another way that VR is being utilized in healthcare is through user experiences and distraction. Users wear the headset to experience a different virtual location while undergoing care interventions or treatments. This has been found to be beneficial for pain management (Ahmadpour et al., 2019), during chemotherapy (Janssen et al., 2022), in wound management (Morris et al., 2010), and following hand surgery (Hoxhallari et al., 2019).

VR is also being used for older adults in hospice and care homes. Users can wear headsets and be transported to different virtual locations as part of reminiscent therapy (Niki et al., 2021). This has been found to improve resident and patient mood, and reduce resident apathy (Chaze et al., 2022; Saredakis et al., 2021).

THE ROLE OF HEALTH AND SOCIAL CARE PROFESSIONALS IN VIRTUAL REALITY

For all the examples discussed, a health or social care professional is required to facilitate the use of the VR. There are challenges and opportunities for the delivery of VR in health and social care environments. Health and social care professionals are the gatekeepers to users accessing and experiencing VR. They require training to implement VR in the healthcare environment, as well as providing technical support, which may be time consuming and costly initially (Markus et al., 2009). The initial purchase of VR equipment can be costly (Sarkar et al., 2021). In addition, the engagement with and buy-in of health professionals is vital for the successful implementation of VR in healthcare settings (Proffitt et al., 2019). Pressures on healthcare professionals, due to workload and time pressures, can be a barrier to implementing VR in busy healthcare settings (Markus et al., 2009). Health and social care professionals' lack of knowledge of the applications of appropriate VR therapies can also be a barrier to implementation (Levac et al., 2017).

Despite VR not being a new technology, it is implemented on a small scale before expansion to more locations. Therefore, access to VR may be limited to where research is being conducted, such as clinics (Sarkar et al., 2021). In addition to this, not all patients are considered suitable for wearing a VR headset, due to vestibular conditions or cognitive issues (Waycott et al., 2022). New research is emerging in relation to improvements to the patient experience and patient outcomes (Ahmadpour et al., 2019; Janssen et al., 2022). Health and social care professionals can be considered "champions" of facilitating innovative, supportive therapies for patients, which can assist in gaining recognition and further utilization of VR (Glegg & Levac, 2018).

Utilizing user-centered design with health and social care professionals and patients in the creation of VR apps is important for professional and patient usability and acceptability (Ashtari et al., 2020). Demonstrating therapeutic effectiveness and reporting direct patient feedback from using VR is beneficial for the encouragement of other users who may have never used VR before (Sarkar et al., 2021). VR is engaging, easily scalable, can enrich the patient experience, creates a prompt for discussion, and can stimulate more connections between staff and patients or other residents (Waycott et al., 2022).

REHABILITATION ROBOTS

Robots are another emerging technology increasingly used in the health and social care sector. Stemming from the word "robota" for "laborer," the term robot refers to a machine that can mimic or automate human behavior. These may behave autonomously, semi-autonomously, or passively. Conceptually, the design of a robot is flexible, and may include soft robots that are warm to hold, robots with exoskeletons, robotic arms, humanoids, and social robots with interpersonal capabilities, and nanorobots that are ~50–100 nm wide.

In healthcare, robots can monitor symptoms and support service engagement in rural populations (Orejana et al., 2015), improve medication adherence (Broadbent et al., 2018), provide companionship and decrease loneliness in older adults (Gasteiger et al., 2021), promote hand hygiene in hospitals (Worlikar et al., 2022), and deliver meals, blood, and urine samples (Law et al., 2021). Importantly, robots can perform tasks repeatedly and without breaks, fatigue, or burnout and can be more engaging than other technologies (e.g., iPads) as they are novel and interactive (Lopez-Samaniego et al., 2014).

Robots have also been found to be well suited for rehabilitation purposes in older adults, including for physical rehabilitation, robot-assisted cognitive rehabilitation, and home-based assistance and support. However, regardless of the application, the goal of robot-assisted rehabilitation is to maximize independence, improve quality of life, and support an older adult in living a life of relative normality. Interest in bionic robot-assisted physical rehabilitation therapy has increased in the past few years, offering a standardized environment, home-based therapy, reduced burden on therapists, and flexibility in adapting the intensity and dose of therapy (Gassert & Dietz, 2018; Qassim & Wan Hasan, 2020). Rehabilitation robots can be categorized as either exoskeletons or end-effector devices (Qassim & Wan Hasan, 2020). Exoskeletons are devices that encapsulate a limb by acting as a splint or support, fitted to the joints of a patient. End-effector devices are applied at one distal point and the joints do not match those of the patient. A patient may, for example, place their hand on the device and perform a pre-specified movement.

Robot-assisted therapies might be applied after an event, such as a stroke or spinal cord injury, to help relearn movements (Gassert & Dietz, 2018), or can be used for conditions like osteoarthritis (McGibbon et al., 2021). For example, McGibbon et al. (2021) conducted a randomized study with 24 middle-aged adults with knee osteoarthritis to test the efficacy of the lower-extremity-assistive Keeogo exoskeleton. Although the exoskeleton did not have any immediate effects on physical performance and activity, it induced cumulative effects, including significantly improving stair climbing time and function, and managing pain. Importantly, the device was well tolerated and safe to use, including for climbing stairs.

Robots have also been successful in delivering cognitive rehabilitation. A systematic review of 99 studies on robot-assisted cognitive rehabilitation found that

FIGURE 8.2   Bomy II (Robocare) home-care assistive robot.

most focused on improving social communication skills, such as imitation and turn-taking (Yuan et al., 2021). However, robots have also been developed to improve memory and concentration, and to alleviate stress (Yuan et al., 2021). Cognitive and psychosocial benefits, like improvements to working memory and reduced anxiety, were reported after older adults with mild cognitive impairment played games on a desktop robot (Lee et al., 2020).

Robots like Care-o-Bot and Bomy II can also provide practical home-based support for older adults. These robots can assist with daily tasks and provide reminders to enhance self-care capabilities (Graf et al., 2009). Bomy II (Figure 8.2) was developed by older adults, carers, dementia experts, researchers, and engineers (Gasteiger et al., 2022). The robot interacted with users via text-to-speech and a pan/tilt touchscreen to provide reminders (e.g., medication, exercise, events/activities, wake-up, and bedtime) and delivered cognitive stimulation games.

THE ROLE OF HEALTH AND SOCIAL CARE PROFESSIONALS IN ROBOTICS

Health and social care professionals may help to increase the acceptance and uptake of robotics, design intuitive and useful technology, and support real-world evaluation efforts. They can act as gatekeepers to the technology, by determining when it is appropriate to use robots, in collaboration with patients and caregivers. They may also help to increase acceptance and encourage uptake of robotics in healthcare, by challenging negative representations of robots.

Health and social care professionals should also be involved in the design process. This is because, practically, robots are not always useful or easy to use, and care professionals have an important role in making them patient centered. Traditional design processes often start by focusing on the product and then searching for a problem it can solve (Law et al., 2019). This sometimes relies on developers and engineers imagining the needs of future users (Vandemeulebroucke et al., 2018; Wang et al., 2019). Alternative design processes, like participatory design and co-design with older adults, informal carers, and health and social care professionals can help to uncover real-life challenges and understand contexts in more detail (Gasteiger et al., 2022; Yuan et al., 2021), ensuring patients are the primary focus. Additionally, they are crucial to avoiding deficit-framing and ensuring that robots are intuitive and complement existing health services.

Health and social care professionals may also support robotics research and evaluation. A systematic review of 23 publications covering socially assistive robots in aged care found that many researchers provided older adults with only one opportunity to interact with a robot, or simply asked them to imagine using it after only seeing it (Vandemeulebroucke et al., 2018). Other studies have simulated home-based environments (Bedaf et al., 2018), rather than having participants test robots in their own homes. Evidently, there is an opportunity for healthcare professionals to bridge the gap between robotics design and real-life evaluation, which is important in helping to understand implementation experiences in the contexts for which robots have been designed, and uncovering real-world barriers to usability (Bajones et al., 2019; Gasteiger et al., 2022; O'Connor, 2021; Zsiga et al., 2017). Furthermore, healthcare professionals can support participant recruitment efforts to ensure that samples represent diverse needs and perspectives.

DIGITAL ACCESS AND AGING

This chapter has discussed several different approaches to the use of health technology, with particular examples related to aging and rehabilitation and the role of the health and social care professional. It is important to consider the use of technology in the context of aging, which can be considered to be another condition within the NASSS framework (Greenhalgh et al., 2017).

The COVID-19 pandemic has been a catalyst for the rise in the use of technology to enhance and access health and social care (Wilson et al., 2021). This has included virtual video appointments, use of applications, and remote monitoring as services rapidly adapted to the need to reduce face-to-face appointments and increase digital and virtual ways for people to access healthcare as an alternative (Kunonga et al., 2021). Access to the internet and the use of technology therefore became essential for accessing healthcare during the pandemic which has continued thereafter (Centre for Ageing Better, 2021a & 2021b).

Age from around 60 years is inversely correlated with the use of technology to access health services (Heponiemi et al., 2022). It is estimated that three million people in the UK do not use the internet, 67% of whom are aged 70 and over (ONS, 2019), demonstrating that age can be one of the factors increasing the risk of digital exclusion. Older adults who do not have access to the internet, the digital skills, or the support to use technology are potentially excluded from the health benefits of access to digital health solutions which could positively affect their health and quality of life, with digital access now being seen as a social determinant for health (Sieck et al., 2021). Research included in this chapter involves participants from a wide range of ages, for example, 64 years to 92 years in Hawley-Hague et al. (2020). This research demonstrates, therefore, that age in and of itself should not be seen as a barrier to using health technology if developed and approached in the right way. Level of digital skills, disability, affordability of technology, such as broadband and support, are important factors to consider, regardless of age (Stone, 2021). Ensuring that older adults are confident in and trust the technology is essential when considering the implementation of new health technology (Wilson et al., 2021) and is something in which health and social care professionals play a key supporting role.

DIGITAL SKILLS

Older adults are often described as "narrow users," meaning that they do not always use the broad range of technology available in their everyday lives (Ofcom, 2018) and thus might be less likely to adopt new health technologies. Broadening technology use by increasing the older adult's digital literacy skills and understanding the needs of older people is essential when considering new technological interventions (Heponiemi et al., 2022). Developing applications and technologies in collaboration with older adults and health professionals means that their needs can be more easily considered and therefore they are more likely to adopt new technologies (Hawley-Hague et al. 2020). Support and training are essential when implementing new technology for use by older adults (Wilson et al., 2021).

SOCIAL DEPRIVATION

Digital exclusion is four times more likely in the poorest households (Lloyds Bank, 2020). Older adults from more socially deprived demographics are less likely to have access to the internet and to use the full range of technology available (ONS, 2019). Data poverty, resulting in insufficient funds to pay for broadband internet to use the internet to access health care services or applications, is a barrier to health technology use (Stone, 2021). Lack of access to Wi-Fi and poor 4G connectivity in more deprived households meant that the use of teleconferencing and exercise apps for those in more deprived households was less feasible (Hawley-Hague et al., 2021).

DISABILITY, PHYSICAL FRAILTY

The aging process means that people are more likely to develop long-term health conditions and disabilities, which could affect everyday function and be a barrier to accessing health technologies. It is therefore essential that new technology, such as health apps aimed at older adults, meet their needs, for example, features which are readily customizable to meet individual needs, such as increasing the font size, and technology which is available across a range of platforms, enabling content to be streamlined to reduce information overload (Wilson et al., 2021). It is important, therefore, that new applications are rigorously tested by older adults in real-world situations to ensure usability (Hawley-Hague et al., 2020).

THE ROLE OF THE HEALTH AND SOCIAL CARE PROFESSIONAL IN ADDRESSING DIGITAL EXCLUSION

When developing and implementing new health and social care technologies, it is crucial that researchers and health and social care professionals address the digital divide to prevent digital exclusion and enable equity of access. To reduce the impact of digital exclusion, it is acknowledged that not everyone will have the skills to use or access to technology. Non-digital solutions and methods to access services are therefore essential, providing person-centered choice of methods of delivery to reduce the impact of digital exclusion (Cheng et al., 2020). Social deprivation, which may result in poor access to Wi-Fi, with cost being a barrier to technology use should be highlighted by the health and social care professionals, for example, by advocating for affordable Wi-Fi and referring to support to access technology (Alkureishi et al., 2021). The narrow technological skills of some patients should be addressed by health and social care professionals when introducing new health technologies, including training (Wilson et al., 2021), which could resolve patients' low self-efficacy in technology use (Fox & Connolly, 2018). Health and social care professionals should also consider how their own attitudes towards technology and its use by older adults can affect uptake. Although older adults may experience digital exclusion, it is important for health professionals not to assume that older adults will not engage with technology as, when they have been involved with the development process, they feel valued and are motivated to use new tools to improve their health.

CONCLUSION

To effectively implement and sustain the use of technology in the support of healthcare delivery, whether it's mHealth, teleconferencing, robotics, virtual reality, or other technologies, there needs to be a recognition of the complex and often implicit, personal processes involved on behalf of the health and social care professional and their key role as gatekeepers to patients and clients.

Person-centered care requires strong relationships between patients and health and social care professionals and allows for shared decision-making. Therefore, technologies need to strengthen this partnership and allow a relationship of compassion and respect. The Health Foundation suggests, when implementing technologies, that we need to "allow for differentiated approaches, taking into account who technology works for and when, and ensuring alternative options are available where feasible" (Horton & Hardy, 2021).

Remote healthcare, using all of these technologies, has both individual and organizational related benefits and challenges for health and social care professionals.

What is important is their confidence and ability to use technology and the subsequent impact this can have on its acceptability from participants. Providing training and leadership when technology is implemented, as well as allowing an approach where the technology can enhance rather than replace in-person services, seems to lead to the best implementation and the greatest success. Without this organizational support and flexibility, there is a risk that care could become less person-centered and, instead, its adoption could be focused more on reducing organizational pressure (e.g., cost/time savings).

All the factors within the theoretical model of health professional resistance by Greenhalgh et al. (2014) can be found when exploring the implementation of our technologies with older adults and those with long-term conditions. This model, alongside innovation diffusion models such NASSS (Greenhalgh et al., 2017) and technology acceptance models such as the TAM (Davis, 1989), help us to consider individual barriers such as confidence, ability, ease of use, and attitudes toward the technology but also how wider organizational structural factors, such as training, leadership, and resources, interact with these attitudes.

Furthermore, access to adequate equipment, specific training, and on-going support and leadership are key to positive attitudes, experiences, and implementation by health and social care professionals, as well as patient satisfaction. The ability to deliver a hybrid approach to technology implementation and to use technology where it enhances healthcare relieves anxiety for health and social care professionals and enables them to better meet the needs of patients. It also ensures that they have clinical control over when it is utilized, as they are more likely to engage if they feel it is not just a cost- or time-saving exercise. This flexible person-centered approach is particularly important if patients or users have sensory or cognitive barriers or if English is not their first language. This approach also helps health and social care professionals overcome concerns about accountability and risk, particularly in terms of patient and user safety and confidentiality.

However, careful planning and resources must be in place to prevent overburdening the workforce. The introduction of a new technology can initially be time consuming and costly and, if the right infrastructure is not in place, could lead to a decline in patient-centered care, rather than enhancing it.

The high costs of some of the technologies we have explored continue to create barriers to access. Very few of the technologies discussed have been evaluated as part of routine delivery within healthcare settings, or developed in collaboration with healthcare providers (Gassert & Dietz, 2018; O'Connor, 2021). Therefore, their feasibility and efficacy when implemented within practice is still unknown. Collaboration between health and social care professionals, gerontologists, engineers, and computer scientists is crucial for evidence-based, useful and intuitive technologies for use by older adults and those with long-term conditions.

When working with older adults and those from deprived communities, the issues around digital exclusion and digital literacy become starker. The role of the health and social care professional can become paramount in helping to navigate these challenges and to bridge some of the gaps related to access. Overall, developing technologies with the intended users, the involvement of both health and social care professionals and a variety of patients and broader stakeholders is paramount to ensure confidence. Providing a robust evidence base with the end users, rather than a convenience sample of more "healthy users," is also key (albeit challenging) and requires engagement with a range of stakeholders and the commitment of health and social care professionals. There are still a large number of apps and healthcare technologies available on the market that lack a sufficiently robust evidence base and that are not developed with those who are expected to adopt them (McGarrigle et al., 2020). Therefore, researchers and developers must continue to engage with patients, users, and the health and social care workforce throughout the development of their innovations, be willing to constantly develop and improve their offer, and to consider their products' wider implementation and diffusion within the workplace. All of this must be done with a consistent focus on keeping the patient at the center of all care.

REFERENCES

Abramsky, H., Kaur, P., Robitaille, M., Taggio, L., Kosemetzky, P.K., Foster, H., Gibson, B.E., Bergeron, M., Jachyra, P. (2018). Patients' perspectives on and experiences of home exercise programmes delivered with a mobile application. *Physiotherapie Canada*, 70(2), 171–178

Aghdam, M.R.F., Vodovnik, A., Hameed, R.A. (2019). Role of telemedicine in multidisciplinary team meetings. *Journal of Pathology Informatics*, 10(1), 35. https://doi.org/10.4103/jpi.jpi_20_19

Ahmadpour, N., Randall, H., Choksi, H., Gao, A., Vaughan, C., Poronnik, P. (2019). Virtual reality interventions for acute and chronic pain management. *The International Journal of Biochemistry and Cell Biology*, 114, 105568. https://doi.org/10.1016/j.biocel.2019.105568

Alkureishi, M.A., Choo, Z., Rahman, A., Ho, K., Benning-Shorb, J., Lenti, G., Velázquez Sánchez, I., Zhu, M., Shah, S.D., Lee, W.W. (2021). Digitally disconnected: Qualitative study of patient perspectives on the digital divide and potential solutions. *JMIR Human Factors*, 8(4), e33364–e33364. https://doi.org/10.2196/33364

Ashtari, N., Bunt, A., McGrenere, Nebeling, M., Chilana, P.K. (2020). Creating augmented and virtual reality applications: Current practices, challenges, and opportinities. In *Proceedings of the 2020 CHI Conference of Human Factors in Computing Systems (CH'20)*. Association for Computing Machinery, New York, pp. 1–12. https://doi.org/10.1145/3313831.3376722

Bajones, M., Fischinger, D., Weiss, A., de la Puente, P., Wolf, D., Vincze, M., Koertner, T., Weninger, M., Papoutsakis, K., Michel, D., Qammaz, A., Panteleris, P., Foukarakis, M., Adami, I., Ioannidi, D., Leonidis, A., Antona, M., Argyros, A., Mayer, P., Panek, P., Eftring, H., Frennert, S. (2019). Results of field trials with a mobile service robot for older adults in 16 private households. *ACM Transactions on Human-Robot Interaction*, 9(2), 1–27. https://doi.org/10.1145/3368554

Bedaf, S., Marti, P., Amirabdollahian, F., de Witte, L. (2018). A multi-perspective evaluation of a service robot for seniors: The voice of different stakeholders. *Disability and Rehabilitation: Assistive Technology*, 13(6), 592–599. https://doi.org/10.1080/17483107.2017.1358300

Bekelis, K., Calnan, D., Simmons, N., MacKenzie, T.A., Kakoulides, G. (2017). Effect of an immersive preoperative virtual reality experience on patient reported outcomes: A randomized controlled trial. *Annals of Surgery*, 265(6), 1068–1073. https://doi.org/10.1097/SLA.0000000000002094

Botella, C., Fernández-Álvarez, J., Guillén, V., García-Palacios, A., Baños, R. (2017). Recent progress in virtual reality exposure therapy for phobias: A systematic review. *Current Psychiatry Reports*, 19(7), 42. https://doi.org/10.1007/s11920-017-0788-4

Broadbent, E., Garrett, J., Jepsen, N., Li Ogilvie, V., Ahn, H.S., Robinson, H., Peri, K., Kerse, N., Rouse, P., Pillai, A., MacDonald, B. (2018). Using robots at home to support patients with chronic obstructive pulmonary disease: Pilot randomized controlled trial. *Journal of Medical Internet Research*, 20(2), e45. https://doi.org/10.2196/jmir.8640

Bunn, F., Byrne, G., Kendall, S. (2004). Telephone consultation and triage: Effects on health care use and patient satisfaction. *Cochrane Database of Systematic Reviews*, (3), Art. No.: CD004180. https://doi.org/10.1002/14651858.CD004180.pub2. Accessed 21 September 2022

Centre for Ageing Better (2021a). Briefing: How has COVID-19 changed the landscape of digital inclusion? https://ageing-better.org.uk/sites/default/files/2021-08/Digital-inclusion-landscape-changes-COV19_0.pdf

Centre for Ageing Better (2021b). COVID-19 and the digital divide: Supporting digital inclusion and skills during the pandemic and beyond. https://ageing-better.org.uk/sites/default/files/2021-07/COVID-19-and-the-digital-divide.pdf

Chaze, F., Hayden, L., Azevedo, A., Kamath, A., Bucko, D., Kashlan, Y., Dube, M., De Paula, J., Jackson, A., Reyna, C., Warren-Norton, K., Dupuis, K., Tsotsos, L. (2022). Virtual reality and well-being in older adults: Results from a pilot implementation of virtual reality in long-term care. *Journal of Rehabilitation and Assistive Technologies Engineering*. https://doi.org/10.1177/20556683211072384

Cheng, C., Elsworth, G.R., Osborne, R.H. (2020). Co-designing ehealth and equity solutions: Application of the ophelia (optimizing health literacy and access) process. *Frontiers in Public Health*, 8, 604401. https://doi.org/10.3389/fpubh.2020.604401

Chesham, R., Malouff, J., Schutte, N. (2018). Meta-analysis of the efficacy of virtual reality exposure therapy for social anxiety. *Behaviour Change*, 35(3), 152–166. https://doi.org/10.1017/bec.2018.15

Cipresso, P., Giglioli, I.A.C., Raya, M.A., Riva,G. (2018). The past, present, and future of virtual and augmented reality research: A network and cluster analysis of the literature. *Frontiers in Psychology*, 9. https://doi.org/10.3389/fpsyg.2018.02086

Cottrell, M.A., Hill, A.J., O'Leary, S.P., Raymer, M.E., Russell, T.G. (2018). Patients are willing to use telehealth for the multidisciplinary management of chronic musculoskeletal conditions: A cross-sectional survey. *Journal of Telemedicine and Telecare*, 24(7), 445–452

Cottrell, M.A., O'Leary, S.P., Raymer, M., Hill, A.J., Comans, T., Russell, T.G. (2019). Does tele-rehabilitation result in inferior clinical outcomes compared with in-person care for the management of chronic musculoskeletal spinal conditions in the tertiary hospital setting? A non-randomised pilot clinical trial. *Journal of Telemedicine and Telecare*. 1357633X19887265

Davis, F.D. (1989). Perceived usefulness, perceived ease of use, and user acceptance of information technology. *MIS Quarterly*, 13(3), 319

Del Rosario, M.B., Redmond, S.J., Lovell, N.H. (2015). Tracking the evolution of smartphone sensing for monitoring human movement. *Sensors, Basel*, 15(8), 18901–18933. https://doi.org/10.3390/s150818901

De Rooij, I.J., van de Port, I.G., Meijer, J.G. (2016). Effect of virtual reality training on balance and gait ability in patients with stroke: Systematic review and meta-analysis. *Physical Therapy*, 96(12), 1905–1918. https://doi.org/10.2522/ptj.20160054

Dulai, R., Shunmugam, S.R., Veasey, R.A., Patel, N.R., Sugihara, C., Furniss, S. (2020). An economic evaluation of an advanced video conferencing system for cardiac multidisciplinary team meetings. *International Journal of Clinical Practice*, 74(9), e13562. https://doi.org/10.1111/ijcp.13562

Fox, G., Connolly, R. (2018). Mobile health technology adoption across generations: Narrowing the digital divide. *Information Systems Journal*, 28, 995–1019. https://doi .org/10.1111/isj.12179

Freitas, J.R.S., Velosa, V.H.S., Abreu, L.T.N., Jardim, R.L., Santos, J.A.V., Campos, P.F., Campos, P.F. (2021). Virtual reality exposure treatment in phobias: A systematic review. *Psychiatry Quarterly*, 92(4), 1685–1710. https://doi.org/10.1007/s11126-021-09935-6

Gagnon, M.P., Ngangue, P., Payne-Gagnon, J., Desmartis, M. (2016). M m-Health adoption by healthcare professionals: A systematic review. *Journal of the American Medical Informatics Association JAMIA*, 23(1), 212–220. https://doi.org/10.1093/jamia/ocv052

Gassert, R., Dietz, V. (2018). Rehabilitation robots for the treatment of sensorimotor deficits: A neurophysiological perspective. *Journal of NeuroEngineering and Rehabilitation*, 15(46). https://doi.org/10.1186/s12984-018-0383-x

Gasteiger, N., Ahn, H.S., Lee, C., Lim, J., Macdonald, B.A., Kim, G.H., Broadbent, E. (2022). Participatory design, development, and testing of assistive health robots with older adults: An international four-year project. *ACM Transactions on Human-Robot Interaction*. https://doi.org/10.1145/3533726

Gasteiger, N., Loveys, K., Law, M., Broadbent, E. (2021). Friends from the future: A scoping review of research into robots and computer agents to combat loneliness in older people. *Clinical Interventions in Aging*, 16, 941–971. https://doi.org/10.2147/CIA.S282709

Glegg, S., Levac, D. (2018). Barriers, facilitators and interventions to support virtual reality implementation in rehabilitation: A scoping review. *PM and R*, 10(11), 1237–1251. https://doi.org/10.1016/j.pmrj.2018.07.004

Graf, B., Reiser, U., Hägele, M., Mauz, K., Klein, P. (2009). *Robotic Home Assistant Care-o-bot® 3—Product Vision and Innovation Platform.* Paper presented at the Proceedings of the IEEE Workshop on Advanced Robotics and Its Social Impacts. Tokyo, Japan

Graffigna, G., Barello, S., Bonanomi, A., Menichetti, J. (2016). The motivating function of healthcare professional in ehealth and mhealth interventions for type 2 diabetes patients and the mediating role of patient engagement. *Journal of Diabetes Research*, 2016, 2974521. https://doi.org/10.1155/2016/2974521

Greenhalgh, T., Swinglehurst, D., Stones, R. (2014). Rethinking resistance to 'big IT': A sociological study of why and when healthcare staff do not use nationally mandated information and communication technologies. *Health Services and Delivery Research*, 39(2), 1–86

Greenhalgh, T., Wherton, J., Papoutsi, C., Lynch, J., Hughes, G., A'Court, C., Hinder, S., Fahy, N., Procter, R., Shaw, S. (2017). Beyond adoption: A new framework for theorizing and evaluating nonadoption, abandonment, and challenges to the scale-up, spread, and sustainability of health and care technologies. *Journal of Medical Internet Research*, 19(11), e367. https://doi.org/10.2196/jmir.8775

Hawley-Hague, H., Gluchowski, A., Lasrado, R., Martinez, E., Akhtar, S., Stanmore, E., Tyson, S. (2022). Opportunities and challenges of delivering remote physiotherapy in the UK during the COVID-19 pandemic: UK wide service evaluation. https://www.csp .org.uk/news/coronavirus/remote-service-delivery-options/national-evaluation-remote -physiotherapy-3

Hawley-Hague, H., Gluchowski, A., Lasrado, R., Martinez, E. Akhtar, S., Stanmore, E., Tyson, S. (2023). Exploring the delivery of remote physiotherapy during the COVID-19 pandemic: UK wide service evaluation. *Physiotherapy Theory and Practice*. https://doi .org/10.1080/09593985.2023.2247069

Hawley-Hague, H., Tacconi, C., Mellone, S., Martinez, E., Chiari, L., Helbostad, J., Todd, C. (2021). One-to-one and group-based teleconferencing for falls rehabilitation: Usability, acceptability, and feasibility study. *JMIR Rehabilitation and Assistive Technologies*, 8(1), e19690. https://doi.org/10.2196/19690

Hawley-Hague, H., Tacconi, C., Mellone, S., Martinez, E., Ford, C., Chiari, L., Helbostad, J., Todd, C. (2020). Smartphone apps to support falls rehabilitation exercise: App development and usability and acceptability study. *JMIR mHealth and uHealth*, 8(9), e15460. https://doi.org/10.2196/15460

Heponiemi, T., Kaihlanen, A., Kouvonen, A., Leemann, L., Taipale, S., Gluschkoff, K. (2022). The role of age and digital competence on the use of online health and social care services: A cross-sectional population-based survey. *Digital Health*, 8, 20552076221074485. https://doi.org/10.1177/20552076221074485

Horton, T., Hardie, T. (2021). *The Health Foundation*. https://www.health.org.uk/news-and -comment/charts-and-infographics/patients-and-machines-does-technology-help-or -hinder-empathy

Hoxhallari, E., Behr, I., Bradshaw, J., Morkos, M., Haan, P., Schaefer, M., Clarkson, J. (2019). Virtual reality improves the patient experience during wide-awake local anesthesia no tourniquet hand surgery: A single-blind, randomized, prospective study. *Plastic and Reconstructive Surgery*, 144(2), 408–414. https://doi.org/10.1097/PRS .0000000000005831

Janssen, A., Fletcher, J., Keep, M., Ahmadpour, N., Rouf, A., Marthick, M., Booth, R. (2022). Experiences of patients undergoing chemotherapy with virtual reality: Mixed methods feasibility study. *JMIR Serious Games*, 10(1), e29579. https://doi.org/10.2196/29579

King, A.C., Hekler, E.B., Grieco, L.A., Winter, S.J., Sheats, J.L., Buman, M.P., Banerjee, B., Robinson, T.N., Cirimele, J. (2013). Harnessing different motivational frames via mobile phones to promote daily physical activity and reduce sedentary behavior in aging adults. *PLOS ONE*, 8. https://doi.org/10.1371/journal.pone.0062613

Kiziltas, D., Celikcan, U. (2018). Knee upan exercise game for standing knee raises by motion capture with RGB-D sensor. In *Smart Tools and Applications for Graphics. Eurographics Italian Chapter Conference*. https://doi.org/10.2312/stag.20181298

Koh, J.S.G., Hill, A.M., Hill, K.D., Etherton-Beer, C., Francis-Coad, J., Bell, E., Bainbridge, L., de Jong, L.D., de Jong, L.D. (2020). Evaluating a novel multifactorial falls prevention activity programme for community-dwelling older people after stroke: A mixed-method feasibility study. *Clinical Interventions in Aging*, 15, 1099. https://doi.org/10 .2147/CIA.S251516

Kothgassner, O.D., Goreis, A., Kafka, J.X., Van Eickels, R.L., Plener, P.L., Felnhofer, A. (2019). Virtual reality exposure therapy for posttraumatic stress disorder (PTSD): A meta-analysis. *European Journal of Psychotraumatology*, 10(1), e1654782. https://doi .org/10.1080/20008198.2019.1654782

Kunonga, T.P., Spiers, G.F., Beyer, F.R., Hanratty, B., Boulton, E., Hall, A., Bower, P., Todd, C., Craig, D. (2021). Effects of digital technologies on older people's access to health and social care: Umbrella review. *Journal of Medical Internet Research*, 23(11), e25887–e25887. https://doi.org/10.2196/25887

Law, M., Ahn, H.S., Broadbent, E., Peri, K., Kerse, N., Topou, E., Gasteiger, N., MacDonald, B.A. (2021). Case studies on the usability, acceptability and functionality of autonomous mobile delivery robots in real-world healthcare settings. *Intelligent Service Robotics*, 14(3), 387–398. https://doi.org/10.1007/s11370-021-00368-5

Law, M., Sutherland, C., Ahn, H.S., MacDonald, B., Peri, K., Johanson, D., Vajsakovic, D., Kerse, N., Broadbent, E. (2019). Developing assistive robots for people with mild cognitive impairment and mild dementia: A qualitative study with older adults and experts in aged care. *BMJ Open*, 9(9), e031937

Lee, E., Kim, B., Kim, H., Kim, S.-H., Chun, M., Park, H., Jeong, J.H., Kim, G.H., Kim, G.H. (2020). Four-week, home-based, robot cognitive intervention for patients with mild cognitive impairment: A pilot randomized controlled trial. *Dement Neurocogn Disord*, 19(3), 96–107

Levac, D., Glegg, S., Colquhoun, H., Miller, P., Noubary, F. (2017). Virtual reality and active videogame-based practice, learning needs, and preferences: A Cross-Canada survey of physical therapists and occupational therapists. *Games for Health Journal*, 6(4), 217–228. https://doi.org/10.1089/g4h.2016.0089

Levinger, P., Hallam, K., Fraser, D., Pile, R., Ardern, C., Moreira, B., Talbot, S. (2017). A novel web-support intervention to promote recovery following anterior cruciate ligament reconstruction: A pilot randomised controlled trial. *Physical Therapy in Sport*, 27, 29–37

Lopez-Samaniego, L., Garcia-Zapirain, B., Mendez-Zorrilla, A. (2014). Memory and accurate processing brain rehabilitation for the elderly: LEGO robot and iPad case study. *Bio-Medical Materials and Engineering*, 24(6), 3549–3556.

Lloyds Bank (2020). *UK Consumer Digital Index 2020. The UK's Largest Study of Transactional, Behaviour and Attitudinal Research Including the New Essential Digital Skills Measure.* https://www.lloydsbank.com/assets/media/pdfs/ banking_with_us/whats-happening/210519- lloyds-cdi-2020-updated-report.pdf

MacNeill, V., Sanders, C., Fitzpatrick, R., Hendy, J., Barlow, J., Knapp, M., Rogers, A., Bardsley, M., Newman, S.P. (2014). Experiences of front-line health professionals in the delivery of telehealth: A qualitative study. *British Journal of General Practice*, 64(624), e401–e407

Malliaras, P., Merolli, M., Williams, C.M., Caneiro, J.P., Haines, T., Barton, C. (2021). 'It's not hands-on therapy, so it's very limited': Telehealth use and views among allied health clinicians during the coronavirus pandemic. *Musculoskeletal Science and Practice*, 102340. https://doi.org/10.1016/j.msksp.2021.102340

Malloy, K.M., Milling, L.S. (2010). The effectiveness of virtual reality distraction for pain reduction: A systematic review. *Clinical Psychology Review*, 30(8), 1011–1018. https://doi.org/10.1016/j.cpr.2010.07.001

Markus, L., Willems, K., Maruna, C., Schmitz, C., Pellino, T., Wish, J., Faucher, L., Schurr, M. (2009). Virtual reality: Feasibility of implementation in a regional burn centre. *Burns*, 35(7), 967–969. https://doi.org/10.1016/j.burns.2009.01.013

Mazza, M., Kammler-Sücker, K., Leménager, T., Keifer, F., Lenz, B. (2021). Virtual reality: A powerful technology to provide novel insight into treatment mechanisms of addiction. *Translational Psychiatry*, 11(1), 617. https://doi.org/10.1038/s41398-021-01739-3

McGarrigle, L., Boulton, E., Todd, C. (2020). Map the apps: A rapid review of digital approaches to support the engagement of older adults in strength and balance exercises. *BMC Geriatrics*, 20(1), 483. https://doi.org/10.1186/s12877-020-01880-6

McGibbon, C., Sexton, A., Jayaraman, A., Deems-Dluhy, S., Fabara, E., Adans-Dester, C., Bonato, P., Marquis, F., Turmel, S., Belzile, E. (2021). Evaluation of a lower-extremity robotic exoskeleton for people with knee osteoarthritis. *Assistive Technology*. https://doi.org/10.1080/10400435.2021.1887400

Meldrum, D., Glennon, A., Herdmen, S., Murray, D., McConn-Walsh, R. (2012). Virtual reality rehabilitation of balance: Assessment of the usability of the Nintendo Wii® Fit Plus. *Disability and Rehabilitation: Assistive Technology*, 7(3), 205–210. https://doi.org/10.3109/17483107.2011.616922

Mellone, S., Tacconi, C., Schwickert, L., Klenk, J., Becker, C., Chiari, L. (2012). Smartphone-based solutions for fall detection and prevention: The FARSEEING approach. *Zeitschrift für Gerontologie und Geriatrie*, 45(8), 722–727. https://doi.org/10.1007/s00391-012-0404-5

Mendes, F., Pompeu, J., Lobo, A., da Silva, K., Oliveira, T., Zomigniani, A., Piemonte, M. (2012). Motor learning, retention and transfer after virtual-reality-based training in Parkinson's disease – effect of motor and cognitive demands of games: A longitudinal, controlled clinical study. *Physiotherapy*, 98(3), 217–223. https://doi.org/10.1016/j.physio.2012.06.001

Meulenberg, C.J.W., de Bruin, E.D., Marusic, U. (2022). A perspective on implementation of technology-driven exergames for adults as telerehabilitation services. *Frontiers in Psychology*, 17(13), 840863. https://doi.org/10.3389/fpsyg.2022.840863

Mirelman, A., Rochester, L., Maidan, I., Del Din, S., Alcock, L., Nieuwhof, F., Rikkert, M.O., Bloem, B.R., Pelosin, E., Avanzino, L., Abbruzzese, G., Dockx, K., Bekkers, E., Giladi, N., Nieuwboer, A., Hausdorff, J.M. (2016). Addition of a non-immersive virtual reality component to treadmill training to reduce fall risk in older adults (V-TIME): A randomised controlled trial. *Lancet*, 388(10050), 1170–1182. https://doi.org/10.1016/S0140-6736(16)31325-3

Morris, L.D., Louw, Q.A., Crous, L.C. (2010). Feasibility and potential effect of a low-cost virtual reality system on reducing pain and anxiety in adult burn injury patients during physiotherapy in a developing country. *Burns : Journal of the International Society for Burn Injuries*, 36(5), 659–664. https://doi.org/10.1016/j.burns.2009.09.005

Muller, I., Kirby, S., Yardley, L. (2015). Understanding patient experiences of self-managing chronic dizziness: A qualitative study of booklet-based vestibular rehabilitation, with or without remote support. *BMJ Open*, 5(5), e007680

Nawaz, A., Skjæret, N., Helbostad, J.L., Vereijken, B., Boulton, E., Svanaes, D. (2016). Usability and acceptability of balance exergames in older adults: A scoping review. *Health Informatics Journal*, 22(4), 911–931. https://doi.org/10.1177/1460458215598638

Niki, K., Yahara, M., Inagaki, M., Takahashi, N., Watanabe, A., Okuda, T., Ueda, M., Iwai, D., Sato, K., Ito, T. (2021). Immersive virtual reality reminiscence reduces anxiety in the oldest-old without causing serious side effects: A single-center, pilot, and randomized crossover study. *Frontiers in Human Neuroscience*, 14, 598161. https://doi.org/10.3389/fnhum.2020.598161

NHS England (2019). *2019 NHS Long Term Plan*. NHS England, London. https://www.longtermplan.nhs.uk/publication/nhs-long-term-plan/. Accessed August 2021

O'Connor, S. (2021). Exoskeletons in nursing and healthcare: A bionic future. *Clinical Nursing Research*, 30(8), 1123–1126. https://doi.org/10.1177/10547738211038365

Odendaal, W.A., Anstey Watkins, J., Leon, N., Goudge, J., Griffiths, F., Tomlinson, M., Daniels, K. (2020). Health workers' perceptions and experiences of using mhealth technologies to deliver primary healthcare services: A qualitative evidence synthesis. *Cochrane Database of Systematic Reviews*, (3), Art. No.: CD011942. https://doi.org/10.1002/14651858.CD011942.pub2. Accessed 21 September 2022

Ofcom (2018). Consumers' experiences in communications markets access and inclusion. https://www.ofcom. org.uk/research-and-data/multi-sectorresearch/accessibility-research/access-andinclusion

Office for National Statistics (2019). Exploring the UK's digital divide. https://www.ons.gov.uk/peoplepopulationandcommunity/householdcharacteristics/homeinternetandsocialmediausage/articles/exploringtheuksdigitaldivide/2019-03-04

Orejana, J.R., MacDonald, B.A., Ahn, H.S., Peri, K., Broadbent, E. (2015). Healthcare robots in homes of rural older adults. In: Tapus, A., André, E., Martin, J.C., Ferland, F., Ammi, M. (eds), *Social Robotics. ICSR 2015. Lecture Notes in Computer Science*, vol. 9388. Springer, Cham. https://doi.org/10.1007/978-3-319-25554-5_51

Piech, J., Czernicki, K. (2021). Virtual reality rehabilitation and exergames–Physical and psychological impact on fall prevention among the elderly–A literature review. *Applied Sciences*, 11(9), 4098. https://doi.org/10.3390/app11094098

Proffitt, R., Glegg, S., Levac, D., Lange, B. (2019). End-user involvement in rehabilitation virtual reality implementation research. *Journal of Enabling Technologies*, 13(2), 92–100. https://doi.org/10.1108/JET-10-2018-0050: 0.1108/JET-10-2018-0050

Qassim, H.M., Wan Hasan, W.Z. (2020). A review on upper limb rehabilitation robots. *Applied Sciences*, 10(19), 6976. https://doi.org/10.3390/app10196976

Ramage, E., Fini, N., Lynch, E., Marsden, D.L., Patterson, A.J., Said, C.M., English, C. (2021). Look before you leap: Interventions supervised via telehealth involving activities in weight-bearing or standing positions for people after stroke—A scoping review. *Physical Therapy*, pzab073. https://doi.org/10.1093/ptj/pzab073

Rowland, S.P., Fitzgerald, J.E., Holme, T., Powell, J., McGregor, A. (2020). What is the clinical value of mhealth for patients? *NPJ Digital Medicine*, 3, 4. https://doi.org/10.1038/s41746-019-0206-x

Rosly, M.M., Rosly, H.M., Davis, G.M., Husain, R., Hasnan, N. (2017). Exergaming for individuals with neurological disability: A systematic review. *Disability and Rehabilitation*, 39(8), 727–735. https://doi.org/10.3109/09638288.2016.1161086

Salisbury, C., Montgomery, A.A., Hollinghurst, S., Hopper, C., Bishop, A., Franchini, A., Kaur, S., Coast, J., Hall, J., Grove, S., Foster, N.E. (2013). Effectiveness of physiodirect telephone assessment and advice services for patients with musculoskeletal problems: Pragmatic randomised controlled trial. *BMJ*, 346, f43. https://doi.org/10.1136/bmj.f43

Saredakis, D., Keage, H.A.D., Corlis, M., Loetscher, T. (2021). Virtual reality intervention to improve apathy in residential aged care: Protocol for a multisite non-randomised controlled trial. *BMJ Open*, 11(2), e046030. https://doi.org/10.1136/bmjopen-2020-046030

Sarkar, U., Lee, J.E., Nguyen, K.H., Lisker, S., Lyles, C.R. (2021). Barriers and facilitators to the implementation of virtual reality as a pain management modality in academic, community, and safety-net settings: Qualitative analysis. *Journal of Medical Internet Research*, 23(9), e26623. https://doi.org/10.2196/26623

Sieck, C.J., Sheon, A., Ancker, J.S., Castek, J., Callahan, B., Siefer, A. (2021). Digital inclusion as a social determinant of health. *NPJ Digital Medicine*, 4(1), 52. https://doi.org/10.1038/s41746-021-00413-8

Skarbez, R., Brooks Jr, F.P., Whitton, M. (2018). A survey of presence and related concepts. *ACM Computing Surveys*, 50(6), 1–39. https://doi.org/10.1145/3134301

Stanmore, E.K., de Jong, D. L., Skelton, D. A., Meekes, W., Mavroeidi, A., Todd, C. (2019) *Perceptions of Exergames for Falls Prevention Among Seniors and Therapists in Assisted Living Facilities. The Gerontological Society of America, Texas Published in the Gerontologist*, vol. 3. Oxford University Press, pp. 59–60. https://academic.oup.com/innovateage/article/3/Supplement_1/S59/5616904

Stanmore, E.K., Mavroeidi, A., De Jong, L.D., Skelton, D.A., Sutton, C.J., Benedetto, V., Munford, L.A., Meekes, W., Bell, V., Todd, C. (2019). The effectiveness and cost-effectiveness of strength and balance exergames to reduce falls risk for people aged 55 years and older in UK assisted living facilities: A multi-centre, cluster randomised controlled trial. *BMC Medicine*, 17(1), 1–14

Stanmore, E., Stubbs, B., Vancampfort, D., de Bruin, E.D., Firth, J. (2017). The effect of active video games on cognitive functioning in clinical and non-clinical populations: A meta-analysis of randomized controlled trials. *Neuroscience and Biobehavioral Reviews*, 78, 34–43. https://doi.org/10.1016/j.neubiorev.2017.04.011

Stone, E. (2021) *Digital Exclusion and Health Inequalities Briefing Paper*. Good Things Foundation. https://www.goodthingsfoundation.org/wp-content/uploads/2021/08/Good-Things-Foundation-2021-%E2%80%93-Digital-Exclusion-and-Health-Inequalities-Briefing-Paper.pdf

Vandemeulebroucke, V., de Casterle, B., Gastmans, C. (2018). How do older adults experience and perceive socially assistive robots in aged care: A systematic review of qualitative evidence. *Aging and Mental Health*, 22(2), 149–167

Van Diest, M., Lamoth, C.J., Stegenga, J., Verkerke, G.J., Postema, K. (2013). Exergaming for balance training of elderly: State of the art and future developments. *Journal of NeuroEngineering and Rehabilitation*, 10, 101. https://doi.org/10.1186/1743-0003-10-10

Vaportzis, E., Clausen, M.G., Gow, A.J. (2017 Oct 04). Older adults perceptions of technology and barriers to interacting with tablet computers: A focus group study. *Frontiers in Psychology*, 8, 1687

Vo, V., Auroy, L., Sarradon-Eck, A. (2019). Patients' perceptions of mhealth apps: Meta-ethnographic review of qualitative studies. *JMIR mHealth and uHealth*, 7(7), e13817. https://mhealth.jmir.org/2019/7/e13817. https://doi.org/10.2196/13817

Wang, S., Bolling, K., Mao, W., Reichstadt, J., Jeste, D., Kim, H., Nebeker, C. (2019). Technology to support aging in place: Older adults' perspectives. *Healthcare*, 7(2), 60. https://doi.org/10.3390/healthcare7020060

Waycott, J., Kelly, R., Baker, S., Neves, B., Thach, K.S., Lederman, R. (2022). The role of staff in facilitating immersive virtual reality for enrichment in aged care: An ethic of care perspective. 1–17. https://doi.org/10.1145/3491102.3501956

Wilson, J., Heinsch, M., Betts, D., Booth, D., Kay-Lambkin, F. (2021). Barriers and facilitators to the use of e-health by older adults: A scoping review. *BMC Public Health*, 21(1), 1–1556. https://doi.org/10.1186/s12889-021-11623-w

Winter, C., Kern, F., Gall, D., Latoschik, M., Pailu, P., Käthner, I. (2021). Immersive virtual reality during gait rehabilitation increases walking speed and motivation: A usability evaluation with healthy participants and patients with multiple sclerosis and stroke. *Journal of NeuroEngineering and Rehabilitation*, 18(68). https://doi.org/10.1186/s12984-021-00848-w

World Health Organisation (2012). Mhealth: New horizons for health through mobile technologies: Based on the findings of the second global survey on ehealth (global observatory for ehealth series, volume 3). *Healthcare Informatics Research*, 18(3), 231

World Health Organization (2021) *Global Strategy on Digital Health 2020–2025*. Geneva. 9789240020924-eng.pdf (who.int)

Worlikar, H., Vyas Vadhiraj, V., Murray, A., O'Connell, J., Connolly, C., Walsh, J.C., O'Keeffe, D.T. (2022). Is it feasible to use a humanoid robot to promote hand hygiene adherence in a hospital setting? *Infection Prevention in Practice*, 4(1), 100188. https://doi.org/10.1016/j.infpip.2021.100188

Wüest, S., Borghese, N.A., Pirovano, M., Mainetti, R., van de Langenberg, R., de Bruin, E.D. (2014). Usability and effects of an exergame-based balance training program. *Games for Health Journal*, 3(2), 106–114. https://doi.org/10.1089/g4h.2013.0093

Yen, H.Y., Chiu, H.L. (2021). Virtual reality exergames for improving older adult's cognition and depression: A systematic review and meta-analysis of randomized control trials. *Journal of the American Medical Directors Association*, 22(5), 995–1002. https://doi.org/10.1016/j.jamda.2021.03.009

Yuan, F., Klavon, E., Liu, Z., Lopez, R.P., Zhao, X. (2021). A systematic review of robotic rehabilitation for cognitive training. *Frontiers in Robotics and AI*, 8. https://doi.org/10.3389/frobt.2021.605715

Zakerabasali, S., Ayyoubzadeh, S.M., Baniasadi, T., Yazdani, A., Abhari, S. (2021 Oct). Mobile health technology and healthcare providers: Systemic barriers to adoption. *Healthcare Informatics Research*, 27(4), 267–278. https://doi.org/10.4258/hir.2021.27.4.267

Zheng, L., Li, G., Wang, X., Yin, H., Jia, Y., Leng, M., Li, H., Chen, L. (2019). Effect of exergames on physical outcomes in frail elderly: A systematic review. *Aging: Clinical and Experimental Research*, 32(11), 2187–2200

Zsiga, K., Tóth, A., Pilissy, T., Péter, O., Dénes, Z., Fazekas, G. (2017). Evaluation of a companion robot based on field tests with single older adults in their homes. *Assistive Technology*, 30(5), 259–266

9 Video Modeling
Opportunities and Ethical Considerations

Jerry K. Hoepner and Katarina L. Haley

INTRODUCTION

Given the advent of smart technologies, which make video recording a ubiquitous tool in rehabilitation contexts, formal and informal use of video modeling is likely to increase. Along with this opportunity come numerous considerations and challenges. Practitioners who employ video modeling should become familiar with its range of applications to assessment and intervention, operational procedures, and medicolegal/ethical policy issues. This chapter will address those three elements.

VIDEO MODELING TECHNIQUES

ASSESSMENT AND DIAGNOSIS

Video modeling approaches often rely on video recordings for initial assessments. Video recordings offer a convenient first step in analyzing the accuracy, complexity, and efficiency of cognitive, linguistic, and motor speech aspects of communication. As such, these recordings can serve essential functions for differential diagnosis and severity estimation. This includes transcription for cognitive/discourse analyses (i.e., coherence, cohesion, topic initiation, topic maintenance, repairs, clarifications, elaborations, interruptions, extralinguistic and paralinguistic behaviors), linguistic content analysis (i.e., lexical and syntactic complexity), as well as perceptual and acoustic analyses of articulation and prosody. Videos can also be an important tool for conducting observational clinical rating tools (e.g., La Trobe Communication Questionnaire, Measure of Skill in Supported Conversation and Measure of Participation in Conversation, OHW Speech Intelligibility-Language Impairment-Cognitive Communicative Impairment Scales, Pragmatic Rating Scale, Profiling Communication Ability in Dementia). Findings from these analyses are a crucial starting point for video modeling interventions.

DOI: 10.1201/9781003272786-9

INTERVENTION APPROACHES

Video modeling has been used for several decades in a variety of contexts and disciplines. While the focus of this chapter is on the use of video modeling in rehabilitation, references to other settings will be included as we discuss potential applications and extensions of the current uses. In the context of acquired neurogenic communication disorders, video modeling is used for persons with cognitive-communication disorders (e.g., traumatic brain injury), language disorders (e.g., aphasia), and speech disorders (e.g., dysarthria, apraxia). Video modeling encompasses a broad range of techniques, including video self-modeling.

There are several ways in which video modeling is used, including video self-modeling (VSM; Buggey and Ogle, 2012; Cream et al., 2010; Haley et al., 2021; Hoepner et al., 2021; Hoepner and Olson, 2018; Lang et al., 2009), video other-modeling (Henry et al., 2018; Hoepner et al., 2015), Comm-COPE-I (Douglas et al., 2019), video feedback (Schmidt et al., 2013); emulation through metaphoric identity mapping (Ylvisaker et al., 2008), self-coaching (Ylvisaker, 2006), conversational coaching, and interpersonal process recall (IPR; Youse and Coelho, 2009). Broadly, it encompasses use of both video and audio cues, such as Collaborative Interpersonal Strategy Building with Audio Reflection (CISBAR; Iwashita, 2022).

In the area of cognition, video modeling is commonly used to increase self-awareness and self-regulation of communication behaviors. For instance, video modeling can help clients recognize what they are doing well and what they can improve upon. The general idea is to increase positive behaviors and decrease less desirable behaviors. Viewing the video provides in-the-moment, tangible evidence of performance. As such, video modeling can help clients reduce interruptions and tangents, provide information more efficiently, determine the right amount of information to share based on their audience, balance talk time by asking questions and listening, and address extralinguistic or paralinguistic elements such as tone of voice and facial affect. Cognitive interventions for impaired self-awareness and self-regulation often rely heavily on partner training and support. Partner training augmented by video modeling focuses on reducing barriers or demands that make it difficult for the client to be successful and increasing use of supports that promote better self-regulation (Douglas et al., 2019; Hoepner et al., 2021; Hoepner and Olson, 2018; McGraw-Hunter et al., 2006; Togher et al., 2010; Rietdijk et al., 2019a). This can include increasing awareness and intentionality about tone of voice, avoiding quizzing and correcting, avoiding requests for information that is already known, achieving more balanced talk time, and promoting listening skills.

Video modeling can also be used as step-by-step guidance for how to perform skilled movements, including those for speech production. Script training is a good example of this type of application. It is a technique where people with aphasia and people with apraxia of speech practice repeating and recalling phrases and monologues they wish to use in their daily life until they can be produced independently, accurately, and with minimal effort. Traditionally, this type of modeling has been completed face-to-face, with a clinician working side by side with the client to guide them through targeted words, phrases, and passages. With the availability of highly portable video recording technology, it is no longer necessary for the modeling to occur in real time. Video modeling can supplement, and sometimes replace, personal training sessions, making learning more efficient and convenient and often increasing practice time dramatically. Script video modeling typically involves watching and listening to a video-recorded speaker model, then speaking in unison with the model, then gradually fading the model until the client is able to produce the script in unison with the model and, eventually, independently (Bilda, 2011; Fridriksson et al., 2012). Video Implemented Script Training for Aphasia is an example of a comprehensive video-based script training method that involves methods for memorization as well as movement guidance (VISTA; Henry et al., 2018). Even when scripts are not fully memorized, video modeling can be used in real time by playing a script on a portable electronic device and producing it in real time while listening through earpieces, using hardwired or Bluetooth connection (Fridriksson et al., 2012).

Most script training programs rely on other-modeling and because the scripts are customized to the needs of each client, the clinician often serves as the model. There is risk that the juxtaposition of the clinician's skilled and effortless productions next to the imperfect attempts of the client may serve as unintentional reminders of what has been lost and how the act of speaking, once taken for granted, has become both errorful and effortful. In programs like Action for Speech and Communication program (ActionSC), the client's best effort is elicited, recorded, and used for self-modeling practice (Haley et al., 2021). Preliminary evidence suggests this type of self-modeling has positive effects on motivation and confidence (Harmon et al., 2018).

Another important application for video modeling is to train family members, friends, and others to be effective communication partners for adults with neurologic communication disorders. Manualized treatments for jointly addressing communication partner training and social communication impairments following traumatic brain injuries (e.g., TBI Express, TBIconneCT) incorporate a combination of psychoeducation, reviewing videos of unsuccessful and successful roleplayed models, and self-modeling (Togher et al., 2010; Rietdijk et al., 2019b). For aphasia, SPPARC (Supporting Partners of People with Aphasia in Relationships & Conversation; Lock, Wilkinson, and Bryan, 2001) and Better Conversations for Aphasia (Beeke et al., 2013) provide a similar manualized framework.

Video modeling is based on the principles that new skills and behaviors can be learned through observation of others, a principle established by anthropologists in the mid-20th century (Bateson, 1936; Williams, 1952). Within the context of communication disorders, targets for cognition would include self-monitoring and self-awareness, language targets would include access to expressive vocabulary and syntax structure and speech targets would include speech articulation and prosody. In other rehabilitation disciplines, such as physical and occupational therapy, video modeling can be used to model transfers, exercises, and functional ADL tasks such as meal preparation or doing laundry.

Kazdin and Smith (1979) established that learners are more likely to acquire modeled skills when the model is perceived as similar to themselves. If the model is perceived to have more status or power than the learner, acquisition of skills is likely to be reduced. This led researchers to shift towards achievable models by others or to self-modeling. Contemporary counseling and coaching approaches align with the same principles by encouraging self-evaluation through observation and outside of the moment. For instance, self-confrontation is a strategy fostered in motivational interviewing, where open-ended questions about one's performance are intended to alter awareness and uncover evidence for self-determined change (Miller and Rollnick, 2012). Self-coaching can be used to identify one's own errors without eliciting oppositional responses or resistance often encountered when a clinician identifies your errors (Ylvisaker, 2006). Likewise, in acceptance and commitment therapy (ACT), defusion is intended to separate one's thoughts from experiences, allowing them to view self as context, taking on the perspective of an independent evaluator of oneself (Hayes et al., 2004). Self-anchored rating scales (De Shazer, 1993) are a core tool in solutions focused brief therapy (SFBT). To use self-anchored rating scales, clients are asked to rate their current performance level and then to identify what it would take to improve their performance by one level (e.g., if a client rates their performance as 6/10, the clinician would ask, "What would it take to reach 7/10?"). Central to use of self-anchored rating scales is the principle of accepting multiple truths, realities, and potential solutions; in other words what we think in the moment may differ from what we observe after the fact (Franklin et al., 1997; Fox, 2012; Fox et al., 2012; Nelson, 2004)

Theoretical underpinnings of video self-modeling are drawn from Bandura's (1986) social cognitive theory, whereby self-beliefs are critical in cognition (attention and retention), motivation, reproduction (i.e., ability to accurately perform the modeled behavior), and behavior. Viewing a video of oneself increases attention more than viewing a video of others (Bandura and Walters, 1963; Fuller and Manning, 1973). Further, VSM reduces any disparity between the person providing the model and themselves (Rosekrans, 1967). In this way, it parallels Vygotsky's (1934) sociocultural theory of development, where models from more knowledgeable others must be approachable and attainable. Again, self-modeling eliminates the gap between learner and the model. Traditionally, VSM has focused on only positive models because it is easier to increase positive behaviors than to reduce errors but that has been expanded to include challenges as well. Generally, a ratio of four positive to one negative self-models is accepted (Dorwick, 1991). Because self-models are achievable, having already been achieved at some point, an emphasis on successes ensures increasing mastery experiences. Viewing errors and/or approximations can establish the need for further practice and motivation for attaining success. Bandura (1997) identified four key sources of self-efficacy beliefs, including: mastery experiences, vicarious experiences, verbal persuasion, and physiological and affective states. Reviewing mastery experiences through video provides corroboration of ability and thus self-efficacy. This enacts self-persuasion regarding capability. Self-models provide repeated opportunities to see oneself succeed. Focused and intentional review of mostly positive self-models ensures review of more positive reflections than may result from their own behavior (Bandura, 1997), while the video itself provides tangible evidence of one's performance. This reduces self-doubt (i.e., "I cannot do it") because the clinician can guide clients to review their successes. It also reduces lack of self-awareness (i.e., "I don't do that") by referring them to review errors and challenges. Video other-modeling relies on vicarious experiences and models. As such, it is crucial for models to be perceived as attainable by our clients. Therefore, a speech-language pathologist, occupational therapist, physical therapist, etc. may not be the best model for attainable sound articulation, for example.

OPERATIONAL PROCEDURES

RECORDING DEVICES

A variety of devices are available for video recording but ease of access and user friendly recording and playback should be a key decision regarding devices. Portable commercial technologies, such as Smartphones and tablets are often the most user friendly. Recording and playback functions are integrated within the same device, allowing clients and families to take the device home and record in their homes or communities. As long as there is not substantial background interference, these devices can produce video and audio recordings that are adequate for video modeling. Smart technologies offer an additional benefit provided by video recording apps such as Coaches Eye™ and Dartfish™. These technologies are especially pertinent to reviewing movement, as they include grid and goniometer functions. Further, there are settings for speed of play adjustments and you can manually progress or reverse footage with a finger swipe or drag. Camcorders and higher-end video cameras provide the ability to capture higher quality video and sound. Many can connect with a wired or wireless microphone, ensuring better audio quality, even within contexts that have background interference. While playback is possible, it is not as intuitive or visually accessible as Smart technologies. Some clinics have indwelling recording systems, such as VALT (Intelligent Video Systems). VALT (https://www.ipivs.com/) ensures privacy and security to meet the most stringent requirements and is capable of full data encryption. Recording quality is good but playback requires a password protected interface with the recording system. This makes immediate playback and review prohibitive; however, the video is automatically recorded to a secure server, which can be linked to electronic medical records. Each of these technologies offer some advantages and disadvantages relevant to video modeling.

RECORDING CONSIDERATIONS AND POLICIES

Once you have selected a recording device, a number of decisions about the device and recording context are necessary. In the case of Smart technologies, the rehabilitation team must decide whether to use institution-owned devices and/or whether to use or allow use of a client's personal devices. Personal devices offer the advantage of being portable to whatever contexts the client experiences. Further, there is less concern about confidentiality from other clients or facility employees. However, personal devices are not without challenges. Inadvertent recording of other clients or individuals within the rehabilitation context can pose a confidentiality risk for bystanders. Further, accessing the video content for the permanent electronic medical record is a challenge. Using institution-owned devices addresses some of these challenges, making access to videos more confidential and facilitating a process for transmitting the video to the electronic medical record. Logistically, this is feasible for video recording activities within the therapy context but more challenging for recording in the client's home and community. Having a repository of devices for loan can allow the client and their families to capture authentic interactions in their lives, often as a part of their homework. Recordings captured at home and in the community can then be reviewed collaboratively in clinical sessions.

INCORPORATING AS A MEDICAL DOCUMENT IN THE EMR

Video recordings are sometimes saved in the medical record to document speech, language, and communication abilities at a given time or the nature of the intervention methods or results. Acoustic and discourse analysis methods can be applied to saved speech samples and allow evaluation of recovery as well as disease progression.

ETHICAL AND LEGAL CONSIDERATIONS

RISKS ASSOCIATED WITH VIDEO MODELING

The aim of video modeling is to increase movement accuracy, self-awareness, and self-regulation. However, a potential consequence of improved self-awareness is negative self-perception and rumination (Hoepner et al., 2021; Hoepner and Olson, 2018). Improved memory function is also associated with better self-awareness and more effective self-reflection (Ownsworth et al., 2019). Unfortunately, stronger working memory, immediate and delayed verbal memory, and verbal fluency are also associated with increased rumination and mood symptoms (Ownsworth et al., 2019). As such, scaffolding memory and self-awareness through tangible and reviewable supports may come with the side effects of rumination and mood symptoms. Furthermore, individuals who have ruminative tendencies are at greater risk of mental health problems. Schmidt et al. (2013) used video feedback to augment verbal feedback within functional tasks, such as meal preparation. The combination reduced errors more than verbal or experiential feedback alone, without related deteriorations in emotional status. Sutton (2016) proposed that certain factors may mediate the likelihood of rumination and its consequences. These include reflective self-development, insight/acceptance, proactivity, and mindfulness. Left unchecked, rumination predicts reduced intervention benefits and increased costs. Mindful reflection fosters improved self-awareness and persistence in the face of performance-related stress or negative feedback (Feldman et al., 2014).

ETHICAL AND LEGAL POLICIES FOR VIDEO MODELING

Multiple healthcare organizations have identified the importance of informed consent for the use of video recording (American Health Information Management Association, 2001; American Medical Association Code of Medical Ethics (AMA),

2003; Joint Commission on Accreditation of Healthcare Organizations (JCAHO), 2009). Clients need to have a clear understanding of the potential risks of video recording to make informed decisions. Butler (2018) identified six categories of potential risks to patients who have been video recorded, including 1) informed consent policies, 2) informed consent procedures, 3) recorded medical errors, 4) secondary use of recordings, 5) collateral patient information, and 6) public trust issues. Although not all of these necessarily apply to videos created for video modeling, most have some relevance. Recorded medical errors and public trust issues are unlikely to occur with video modeling. Comprehension of informed consent policies and procedures has substantial relevance to video modeling. When working with individuals with language and cognitive impairment, we need to adapt consent forms and use communication supports to ensure their comprehension (Jayes and Palmer, 2014; Kagan and Kimelman, 1995; Palmer and Patterson, 2011). Zuscak et al. (2015) developed a three-step framework for legal and clinical decision-making for persons with communication impairments, which was adapted and modified by Kagan et al. (2020) to ensure active engagement and communicatively supported decisions. This includes ensuring understanding of relevant information delivered in spoken and written form, comprehending the consequences, engaging in reasoning (weighing options and alternatives), and communicating a choice through preferred modality (e.g., verbal, gesture, writing, pointing to words, pictographic illustrations, or drawing). In many cases, it requires that we inform parents and/or caregivers so that they can make an informed decision (proxy consent) for their loved one. Deciding when to conduct informed consent for video recording is also an important decision. While some organizations include video consent within the registration process, along with general release forms, many argue that it should be conducted in the context of initiating video recordings (AMA, 2003; JCAHO, 2009). Secondary use of videos by our clients is not a concern unless there is collateral patient information included. Any secondary use by clinicians for use in publications or presentations must be explicitly requested in a separate consent. Videos that contain collateral patient information are typically best handled by immediate deletion. A sample policy for video modeling is included in Figure 9.1.

The American Health Information Management Association (AHIMA, 2010) developed a sample consent form for clinical photography, video taping, audiotaping, and other multimedia imaging of patients. The AHIMA cautions that customized organization-specific policies and procedures and modifications to this sample are necessary. The sample is available at https://library.ahima.org/doc?oid=99416#.Y4Y _0xTMKM8. The Profiling Communication Ability in Dementia (P-CAD) includes video consent as a part of their examination booklet (Dooley et al., 2022). It provides a nice model for eliciting client consent, proxy consent, communication partner consent, and clinician documentation of consent. Figure 9.2 provides an example of a communicatively accessible consent for video recording. This example was developed drawing upon principles of accessible consent (Jayes and Palmer, 2014; Kagan and Kimelman, 1995; Palmer and Patterson, 2011) and the SOFIA trial resource page   (https://city.figshare.com/collections/SOlution_Focused_brief_therapy_In _post-stroke_Aphasia_SOFIA_feasibility_trial/4491122).

Policy for use of video capture for therapeutic intervention

This policy and statements therein include guidance for the use of video capture as an intervention modality. This applies to video capture for immediate and delayed review in the case of video self-modeling, video other modeling, and related treatment approaches. Failure to comply with the procedures outlined here may result in disciplinary action. Compliance with confidentiality standards and medical records require the following actions:

1) Any video captured within a clinical session, collected for the purpose of immediate review, must be deleted prior to the end of that session.

2) Note that the video capture device (i.e., iPad, tablet, iPod, iPhone, digital video camera, or related devices) must be placed in airplane mode or disconnected from the wireless system (i.e., WiFi disabled). This will ensure that captured videos do not reach the cloud, where data could be retrieved at a later time.

3) Any video captured within a session, collected for the purpose of immediate and delayed (i.e., at a later session) review, must be downloaded manually from the device via cable (i.e., not uploaded through a cloud, dropbox, or related function). This action must take immediately following completion of that session. The clinician is not allowed to take the video capture device to any other patient/client rooms or sessions. Videos must be stored on a secured server and/or electronic medical record system.

4) Any video, which is collected and saved for later review, is at that point considered a part of that patient/client's medical record and thus is held to the standards of other medical records.

5) Video collected for use with a specific client may not be used with other clients.

6) Any captured video that inadvertently or intentionally includes identifiable evidence of other patients/clients and family, must be deleted prior to the end of that session.

7) Any video captured by the patient/client and/or family and other visitors may be used under the discretion of that patient/family member or visitor, UNLESS the captured video inadvertently or intentionally includes identifiable evidence of other patients/clients, family, or visitors. If a clinician is aware of such evidence, they are required to request deletion of the file and report any non-compliance to _____.

8) Electronic sharing of any captured video, which is downloaded and stored on a secure server and/or electronic medical record, is prohibited.

9) Refer to policies on photographic and video consent for further details.

10) Video capture by staff (i.e., including paid staff, students, or unpaid volunteers) must be collected using institutionally owned and compliant devices. Video capture using personal devices is prohibited.

This policy is prepared for the use of video capture for the use of therapeutic intervention but excludes electronic capture of other clinical/medical procedures (e.g., video fluoroscopic examination of swallowing, flexible endoscopic examination of swallowing, stroboscopic examination of laryngeal function). It may include video capture for the use of therapeutic interventions such as: gait analysis, ADLs (excluding showering and dressing), iADLs, and interactional communications.

FIGURE 9.1   Sample policy for video capture.

Consent for video capture for video modeling therapeutic intervention

Please check each box if you agree:

	My therapist has **explained** the purpose of the **video modeling** approach. I have read information about the approach.	☐
	My therapist has explained the **potential risks** associated with **video modeling**.	☐
	I have had the opportunity to **ask questions** about the approach.	☐
	I understand that video modeling means: • **Recording videos** of myself and partners • **Reviewing videos** of myself • **Making judgments** about my own performance	☐
	I understand that videos will be added to my **medical records**.	☐
	I understand that the videos will be **kept safe** and **confidential**.	☐
	I consent to: • Sharing video recorded interactions for therapy. • Recording of interactions during therapy.	☐
	Videos may be used: 1. Only for **therapy** 2. For **teaching** and **training** (e.g., of students and practicing therapists) 3. For **presentations** (e.g., at conferences). I understand that therapists will **check first** before using videos in presentations.	1. ☐ 2. ☐ 3. ☐

_____   _____   _____

Patient   Patient Signature   Date

_____   _____   _____

Witness   Witness Signature   Date

FIGURE 9.2   Communicatively accessible consent.

Icons in the consent form are from Microsoft icons in Microsoft Word.

AHIMA (2001) produced guidelines regarding the use of photos, video, and audio recordings, and other multimedia imaging. These guidelines specify that all photographs, video, and audio must be identified with the patient's name, medical record number, account number, and admission information. Once a part of the medical record and legal health record, it is subject to confidentiality and protected health information policies. Additionally, these records are subject to organizational record retention policies and procedures. Photographs, videos, and audio captured on the organization's devices must adhere to these same policies for confidentiality and retention. The use of personal devices is prohibited. All of these policies must comply with the Health Information Portability and Accountability Act (HIPPA, 1996) privacy rule (Section II).

CONCLUSIONS

Video modeling is an increasingly viable approach to augment other rehabilitation interventions. Several positive outcomes are associated with video modeling, as well as some potential risks. Risks associated with video modeling create ethical and legal considerations, which must be addressed. Outside of research and policies directed toward physicians and health information management, very little direction is available to rehabilitation practitioners regarding the use, handling, and storage of video recordings. Much of the existing evidence is directed toward recordings of medical procedures (e.g., surgeries) and telecommunications. Drawing upon policies from medical and health information management associations, this chapter attempted to generate sample policies and consent forms, which are consistent with best practices. Equipped with policies and consent, rehabilitation practitioners will be better positioned to deliver video modeling interventions in an ethical and legal manner.

REFERENCES

AHIMA, 2010. Sample consent for clinical photography, videotaping, audiotaping, and other multimedia imaging of patients. *Journal of AHIMA*. https://library.ahima.org/doc?oid=99416#.Y4Y_0xTMKM8.

American Health Information Management Association, 2001. Practice brief. Patient photography, videotaping, and other imaging (updated). *Journal of American Health Information Management Association*, 72(6), pp. 64M–64Q.

American Medical Association, 2003. Code of medical ethics. Opinion 5.046 – filming patients for educational purposes. (based on the CEJA report 12-A-03). www.ama-assn.

Bandura, A., 1986. *Social Foundations of Thought and Action: A Social Cognitive Theory.* Prentice-Hall.

Bandura, A., 1997. *Self-Efficacy: The Exercise of Control.* New York, NY: Worth Publishers.

Bandura, A. and Walters, R.H., 1963. *Social Learning and Imitation.* New York: Holt, Rinehart and Winston.

Bateson, G.N., 1936. *A Survey of the Problems Suggested by a Composite Picture of the Culture of a New Guinea Tribe Drawn from Three Points of View*. Cambridge: Cambridge University Press.

Beeke, S., Sirman, N., Beckley, F., Maxim, J., Edwards, S., Swinburn, K. and Best, W., 2013. Better conversations with aphasia: An e-learning resource. https://extend.ucl.ac.uk/.

Bilda, K., 2011. Video-based conversational script training for aphasia: A therapy study. *Aphasiology, 25*(2), pp. 191–201. https://doi.org/10.1080/02687031003798254.

Buggey, T. and Ogle, L., 2012. Video self-modeling. *Psychology in the Schools, 49*(1), pp. 52–70. https://doi.org/10.1002/pits.20618.

Butler, D.J., 2018. A review of published guidance for video recording in medical education. *Families, Systems, and Health, 36*(1), p. 4.

Cream, A., O'Brian, S., Jones, M., Block, S., Harrison, E., Lincoln, M., ... Onslow, M., 2010. Randomized controlled trial of video self-modeling following speech restructuring treatment for stuttering, *Journal of Speech Language Hearing Research 53*(4). 887–897. https://doi.org/10.1044/1092-4388(2009/09-0080).

De Shazer, S., 1993. Creative misunderstanding: There is no escape from language. In S. G. Gilligan & R. Price (Eds.), *Therapeutic conversations* (pp. 81–94). New York, NY: Norton.

Dooley, S., Hopper, T. and Walshe, M., 2022. *Profiling Communication Ability in Dementia (P-CAD)*. Dublin: O'Brien Press LTD.

Douglas, J.M., Knox, L., De Maio, C., Bridge, H., Drummond, M., and Whiteoak, J., 2019. Effectiveness of communication-specific coping intervention for adults with traumatic brain injury: Preliminary results. *Neuropsychological Rehabilitation, 29*(1), pp. 73–91. https://doi.org/10.1080/09602011.2016.1259114.

Dowrick, P.W., 1991. *Practical Guide to Using Video in the Behavioral Sciences*. John Wiley & Sons.

Feldman, G., Dunn, E., Stemke, C., Bell, K. and Greeson, J., 2014. Mindfulness and rumination as predictors of persistence with a distress tolerance task. *Personality and Individual Differences, 56*, pp. 154–158.

Fox, L.E., 2012. AAC collaboration using the self-anchored rating scales (SARS): An aphasia case study. *Perspectives on Augmentative and Alternative Communication, 21*(4), pp. 136–143.

Fox, L.E., Andrews, M.A. and Andrews, J., 2012. Self-anchored rating scales: Creating partnerships for post-aphasia change. *Perspectives on Neurophysiology and Neurogenic Speech and Language Disorders, 22*(1), pp. 18–27.

Franklin, C., Corcoran, J., Nowicki, J. and Streeter, C., 1997. Using client self-anchored scales to measure outcomes in solution-focused therapy. *Journal of Systemic Therapies, 16*(3), p. 246.

Fridriksson, J., Hubbard, H.I., Hudspeth, S.G., Holland, A.L., Bonilha, L., Fromm, D. and Rorden, C., 2012. Speech entrainment enables patients with Broca's aphasia to produce fluent speech. *Brain, 135*(12), pp. 3815–3829.

Fuller, F.F. and Manning, B.A., 1973. Self-confrontation reviewed: A conceptualization for video playback in teacher education. *Review of Educational Research, 43*(4), pp. 469–528.

Haley, K.L., Cunningham, K.T., Kim, I. and Shafer, J.S., 2021. Autonomy-supportive treatment for acquired apraxia of speech: Feasibility and therapeutic effect. *Aphasiology, 35*(4), pp. 539–559.

Harmon, T.G., Hardy, L. and Haley, K.L., 2018. Proactive social validation of methods and procedures used for training speech production in aphasia. *Aphasiology, 32*(8), pp. 922–943.

Hayes, S.C., Strosahl, K.D., Luoma, J., Smith, A.A. and Wilson, K.G., 2004. ACT case formulation. In: *A Practical Guide to Acceptance and Commitment Therapy*, pp. 59–73. Boston, MA: Springer.

Health Insurance Portability and Accountability Act, 1320d. P.L. 1996.104–191, 42 U.S.C.

Henry, M.L., Hubbard, H.I., Grasso, S.M., Mandelli, M.L., Wilson, S.M., Sathishkumar, MT. ... and Gorno-Tempini, M.L., 2018. Retraining speech production and fluency in nonfluent/agrammatic primary progressive aphasia. *Brain*, *141*(6), 1799–1814.

Hoepner, J., Sell, L. and Kooiman, H., 2015. Case study of partner training in corticobasal degeneration. *Journal of Interactional Research in Communication Disorders*, *6*(2), p. 157.

Hoepner, J.K., Sievert, A. and Guenther, K., 2021. Joint VSM for persons with traumatic brain injuries and their partners: A case series. ICCDC special issue of *American Journal of Speech-Language Pathology*, *30*(2S), pp. 863–882.

Hoepner, J.K. and Olson, S.E., 2018. Joint video self-modeling as a conversational intervention for an individual with a traumatic brain injury and his everyday partner: A pilot investigation. *Clinical Archives of Communication Disorders*, *3*(1), pp. 22–41. http://doi.org/10.21849/cacd.2018.00262.

Iwashita, H., 2022. *A Single Case Experimental Design Investigating Collaborative Interpersonal Strategy Building with Audio Reflection (CISBAR) for Improving Social Communication after Acquired Brain Injury*. Doctoral dissertation. University of Oregon.

Jayes, M. and Palmer, R., 2014. Initial evaluation of the consent support tool: A structured procedure to facilitate the inclusion and engagement of people with aphasia in the informed consent process. *International Journal of Speech-Language Pathology*, *16*(2), pp. 159–168.

Joint Commission on Accreditation of Healthcare Organizations, 2009. Standards and elements of performance. RI. 01.03.03. The hospital honors the patient's right to give or withhold informed consent to produce or use recordings, films or other images for purposes other than his or her care. http://www.jointcommission.org/assets/1/6/2009_CLASRelatedStandardsAHC.pdf.

Kagan, A., Shumway, E. and MacDonald, S., 2020. Assumptions about decision-making capacity and aphasia: Ethical implications and impact. *Seminars in Speech and Language*, *41*(3), pp. 221–231. Thieme Medical Publishers.

Kagan, A. and Kimelman, M.D., 1995. Informed consent in aphasia research: Myth or reality. *Clinical Aphasiology*, *23*, pp. 65–75.

Kazdin, A.E. and Smith, G.A., 1979. Covert conditioning: A review and evaluation. *Advances in Behaviour Research and Therapy*, *2*(2), pp. 57–98.

Lang, R., Shogren, K.A., Machalicek, W., Rispoli, M., O'Reilly, M., Baker, S. and Regester, A., 2009. Video self-modeling to teach classroom rules to two students with Asperger's. *Research in Autism Spectrum Disorders*, *3*(2), pp. 483–488.

Lock, S., Wilkinson, R., Bryan, K., Maxim, J., Edmundson, A., Bruce, C. and Moir, D., 2001. Supporting partners of people with aphasia in relationships and conversation (SPPARC). *International Journal of Language & Communication Disorders*, *36*(supl), pp. 25–30.

McGraw-Hunter, M., Faw, G.D. and Davis, P.K., 2006. The use of video self-modeling and feedback to teach cooking skills to individuals with traumatic brain injury: A pilot study. *Brain Injury*, *20*(10), pp. 1061–1068.

Miller, W.R. and Rollnick, S., 2012. *Motivational Interviewing: HELPING People Change*. Guilford Press.

Nelson, L.J., 2004. Clinical issues: Using self-anchored rating scales in family-centered treatment. *Perspectives on Language Learning and Education*, *11*(1), pp. 14–17.

Ownsworth, T., Gooding, K. and Beadle, E., 2019. Self-focused processing after severe traumatic brain injury: Relationship to neurocognitive functioning and mood symptoms. *British Journal of Clinical Psychology*, *58*(1), pp. 35–50.

Palmer, R. and Paterson, G., 2011. One size does not fit all: Obtaining informed consent from people with aphasia. *ACNR*, *11*(2), pp. 30–31.

Rietdijk, R., Power, E., Brunner, M. and Togher, L., 2019a. A single case experimental design study on improving social communication skills after traumatic brain injury using communication partner telehealth training. *Brain Injury*, *33*(1), pp. 94–104.

Rietdijk, R., Power, E., Brunner, M., Attard, M., McDonald, S., Tate, R., & Togher, L., 2019b. TBIconneCT clinician manual: Connecting people living with traumatic brain injury to conversation training. Sydney, Australia: Australian Society for the Study of Brain Impairment.

Rosekrans, M.A., 1967. Imitation in children as a function of perceived similarity to a social model and vicarious reinforcement. *Journal of Personality and Social Psychology*, *7*(3p1), p. 307.

Schmidt, J., Fleming, J., Ownsworth, T. and Lannin, N.A., 2013. Video feedback on functional task performance improves self awareness after traumatic brain injury: A randomized controlled trial. *Neurorehabilitation and Neural Repair*, *27*(4), pp. 316–324. https://doi.org/10.1177/1545968312469838.

Sutton, A., 2016. Measuring the effects of self-awareness: Construction of the self-awareness outcomes questionnaire. *Europe's Journal of Psychology*, *12*(4), p. 645.

Togher, L., McDonald, S., Tate, R., Power, E. and Rietdijk, R., 2010. *TBI Express: For People with TBI and Their Everyday Conversational Partners*. The Australian Society for the Study of Brain Impairment (ASSIBI).

VALT (Video Audio Learning Tool). *Intelligent Video Solutions*. N53W24747 S Corporate Circle, Sussex, WI 53089. https://www.ipivs.com/.

Vygotsky, L.S., 1934. Thinking and speech: psychological research. Moscow: Labyrinth.

Williams, R.M., 1952. Experimental designs for serially correlated observations. *Biometrika*, *39*(1/2), pp. 151–167.

Ylvisaker, M., 2006. Self-coaching: A context-sensitive, person-centred approach to social communication after traumatic brain injury. *Brain Impairment*, *7*(3), pp. 246–258. https://doi.org/10.1375/brim.7.3.246.

Ylvisaker, M., Mcpherson, K., Kayes, N. and Pellett, E., 2008. Metaphoric identity mapping: Facilitating goal setting and engagement in rehabilitation after traumatic brain injury. *Neuropsychological Rehabilitation*, *18*(5–6), 713–741.

Youse, K.M. and Coelho, C.A., 2009. Treating underlying attention deficits as a means for improving conversational discourse in individuals with closed head injury: A preliminary study. *NeuroRehabilitation*, *24*(4), pp. 355–364.

Zuscak S.J., Peisah C. and Ferguson A., 2015. A collaborative approach to supporting communication in the assessment of decision making capacity. *Disability and Rehabilitation*, *12*, pp. 1–8. https://doi.org/10.3109/09638288.2015.1092176.

10 The Explosion of Technology in Pediatric Rehabilitation
A Call to Use the F-Word Lens

Anna McCormick, Hana Alazem, Elaine Biddiss,
Joanna Butchart, Deborah Gaebler Spira,
Jan Willem Gorter, and Adam Scheinberg

INTRODUCTION

The use of technology in rehabilitation is exploding internationally (Reyes et al., 2020). Innovative technologies are increasing hope for improved functional outcomes and participation. In tandem, a new lens is being applied to pediatric disability. The F-words in Child Development (Function, Family, Fitness, Fun, Friends, and Future), built upon the World Health Organization's International Classification of Functioning, Disability and Health (ICF) framework (Rosenbaum and Gorter, 2012), are attracting attention around the world (Soper et al., 2020a,b; Nguyen et al., 2019; Soper et al., 2019). It is time to apply this universal, holistic, interactive model to reflect on the development, use, and assessment of the multitude of devices which should be purposefully designed to meet the self-identified needs of those with disabilities (Mensah-Gourmel et al., 2022), and have positive impacts on children living with disabilities and their families.

In 2001, the WHO launched the ICF framework, to provide a standardized global metric for the impact of medical conditions on an individual and populations, which builds on and complements the International Classification of Diseases (ICD) and the International Classification of Health Interventions (ICHI) (Stucki et al., 2008). Given that the ICF extends beyond the individual and incorporates environmental and personal factors and their interactions, it gives a more holistic view of disability and disease at the patient and population levels (Stucki et al., 2008; Stucki, 2009). The ICF looks primarily at function of an individual: of the body, in social and familial activities, as a member of society, and within the environment in relation to assistance or barriers posed. It uses a scale to rank the severity of impact for each of the categories as a way to show progression over time or with specific interventions (Stucki et al., 2008). The use of this tool can assist in the planning and execution of treatment strategies to optimize quality of life, even if that does not include what would be considered standard functional gains pertaining to the severity of the specific disorder (Stucki, 2009). Standardization of common data elements and descriptions allows researchers around the world to understand and classify their patients within a global milieu.

As an extension to the ICF framework, researchers at CanChild have developed a set of criteria for the assessment of disability called the F-words (Rosenbaum and Gorter, 2012; Soper et al., 2020a,b; Nguyen et al., 2019; Soper et al., 2019). These key words – function, family, fitness, fun, friends, and future – provide practitioners with concrete examples of ways in which a treatment or technology can have an impact on a patient's life. By viewing them from many perspectives, the F-words aim to look beyond simply functional or medical gains and assess how the intervention can benefit the unique individual socially, emotionally, and within the family dynamic. With this assessment method, we endeavor to focus on the complex and multifaceted person, rather than simply the physical functional aspects of their condition.

The ICF and F-words classifications have overlapping elements with slightly different lenses. Although the ICF is broadly applicable across all age groups, with complementary child and youth versions (Simeonsson et al., 2003), it is largely focused on the function of a disabled adult in society. The F-words, on the other hand, have a more pediatric focus and look at children, youth, and the surrounding family unit. It allows plain language discussions about priorities at different ages and stages of development. Both scales are invaluable as they shift the focus of the provider from helping someone "get better" to helping them "live better."

Through decades of device development and evaluation, professionals have been trained to ask a series of questions: Does the new technology meet the needs of the patient? Is it easy to use and affordable? Does it enhance function or facilitate ease of care? Equally important are questions to ensure that there are no unintended consequences or harm related to the use of this technology. Symposia at the American Academy for Cerebral Palsy and Developmental Medicine (AACPDM) (McCormick et al., 2019a), The American Congress of Rehabilitation Medicine (ACRM) (McCormick et al., 2019b) and the European Academy of Childhood Disability (EACD) (McCormick et al., 2019) provide fora for analysis and group feedback.

In this chapter, we illustrate the use of the ICF/F-words framework to address the questions posed above, while adding depth to the understanding of current and future potential for impact. The framework challenges the reader to look beyond the here-and-now, beyond basic function, and beyond traditional biomedical and rehabilitation assessments by asking bigger, broader questions: Are the devices fun to use? Do they enhance physical and mental fitness? Do they involve and impact families? Can they enhance and nurture friendships? Are they likely to enhance the future function of individuals as adults? We assess their relevance and real-life impact in the natural environments of the child at home, at school, and in the community.

FIGURE 10.1   Therapeutic mixed-reality games (Bootle Blast©).

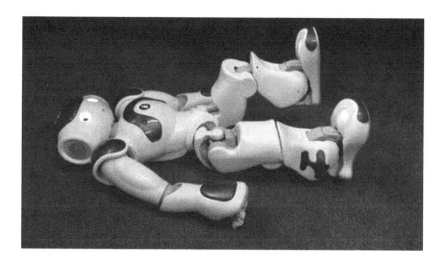

FIGURE 10.2   Socially assistive robot (NAO).

This analysis focuses on three examples that represent innovative, evolving areas of technological development: Therapeutic Virtual Reality Games (Bootle Blast©, Figure 10.1), the socially assistive robot (NAO, Figure 10.2), and robotic walkers (SoloWalk, Figure 10.3), (Trexo Plus, Figure 10.4). These devices have been assessed at the AACPDM and ACRM as offering new hope (McCormick et al., 2019a,b) for assistance in rehabilitation therapy and improvement in functional outcomes.

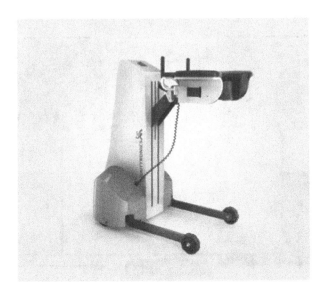

FIGURE 10.3   Robotic walker (SoloWalk).

FIGURE 10.4   Robotic walker (Trexo Plus).

THERAPEUTIC VIRTUAL REALITY GAMES: BOOTLE BLAST©

THE TECHNOLOGY

Children and youth have co-created therapeutic video games with engineers, clinicians, game developers, kids' media specialists, and a composer based at the

Possibility Engineering and Research Lab (PEARL) at Holland Bloorview Kids Rehabilitation Hospital in Toronto, Canada. This team is guided by contemporary theories of motivation and best practices in gamification to promote engagement in therapy gaming (Biddiss et al., 2019, Biddis et al 2021).

Defining features of Bootle Blast© include the ability to target gross and fine motor movement through a mixed-reality play experience that tracks and rewards activities, such as reach, bilateral coordination, and manipulation of real-world objects, including musical instruments and building blocks. Bootle Blast© can be configured to each child's abilities and therapy goals and the games progress in difficulty to encourage feelings of mastery and growth in competence. Practice goals can be personalized and incentivized by rewards built into the game narrative. The games provide opportunities for both competitive and cooperative multiplayer activities, calibrated to individual abilities to even the playing field. Social play, alongside therapist and parent support, aims to facilitate feelings of relatedness to sustain engagement.

REHABILITATION EVALUATION

A 13-year-old boy with unilateral spastic cerebral palsy (USCP), with good manual abilities (Manual Ability Classification System (MACS) Level I), achieved his self-identified practice goal of 15 minutes of active game play, three times per week, in ten weeks of the 12-week trial. In total, the participant achieved 9.4 hours of active game play. Improvements were documented in function and body structure improvements, including a 15 degree increase in active range of motion of the wrist. There were activity gains of +5 units on the Assisting Hand Assessment, a measure of bimanual hand use, and gains in self-identified goal areas (e.g., tying shoelaces, using a computer with hemiplegic hand) with improvements of +4 units in satisfaction and +3 units in performance on the Canadian Occupational Performance Measure (COPM) for the primary goal (tying shoelaces). Qualitative post-intervention interviews and surveys indicated high and sustained levels of affective, behavioral, and cognitive engagement. The games are currently being piloted at home with 15 children with USCP in Costa Rica.

F-WORD LENS

Bootle Blast©, coupled with growing evidence of successful therapeutic virtual reality (VR) use in individuals with disability (Ravi et al., 2017; Gong et al., 2017; Chen et al., 2018; Porras et al., 2018), gives reason for optimism regarding virtual therapeutic options. Improvements are highlighted by the use of standardized scales for upper extremity function, through patient-reported outcome measures, and interviews with children and their families. Use of real-life activities and integration of common objects utilized for play seek to enable carry-over into the child's natural environment.

The elements of play, storyline, imaginative characters, and achievement levels mirror mainstream entertainment gaming. The qualitative data assessing affect and behavior support the literature and are evidenced by decades of video game purchase and use: video games have high appeal for children with and without disabilities (Ravi et al., 2017; Shikako-Thomas et al., 2015; Bonnechère et al., 2016). For many, they are fun!

From a social perspective, these games have modes to involve families and friends. VR games can also be adjusted to provide access for many levels of ability. In this device, a single-switch accessible version is in development to provide activities feasible for individuals with more extensive motor involvement, namely those in Gross Motor Function Classification System (GMFCS) levels IV and V, answering a call to break down barriers and enable the benefits of participation for all (Anaby et al., 2020; Handler et al., 2019).

Beyond concrete family involvement, therapy games can also decrease family stress by shifting some of the responsibility for motivating home practice from parents to technology. Along with savings in time and money as a result of fewer trips to a care facility, as well as decreasing the risk of infectious contact (an issue of heightened importance during the COVID-19 pandemic crisis) (Murphy, 2020), these findings emphasize the potential real-world impact on family life.

Fitness has been a focus of other virtual therapeutic interventions and gaming (Knights et al., 2016; Lewis et al., 2011; Sevick et al., 2016). In the design of virtual reality (VR) testing facilities, air conditioning and good ventilation are recommended, as keen participants "work up a sweat." With clever design that links game rewards to desired outcomes (e.g., reaching a target heart rate), VR may support fitness goals as well.

As we consider fitness, fun and family, it behooves us to consider the possibility of unintended consequences. Will increasing use of VR add to the screen time that parents have been asked to limit, causing more family friction? Will it further socially distance individuals who may already feel isolated? Concerns have also been raised regarding the risk of online bullying if gaming with strangers? As with most developments, users will need to balance the pros and cons of the use of these tools.

In the future, a user envisions optimizing function through increased access to convenient, enjoyable, customized, therapeutic intervention across the lifespan. Children, youth, and families would see a future with more options for therapeutic intervention, less time spent in waiting rooms and in care facilities, less time spent on waiting lists, or advocating for scarce services, allowing more time for family, friends, fitness, and fun. One can also imagine the possibility for vocational implications, an area where there is much need for growth. Ongoing co-design holds potential opportunities for employment, changing financial trajectories for those living with disabilities and therefore offering new hope.

SOCIALLY ASSISTIVE ROBOT: NAO

THE TECHNOLOGY

A socially assistive robot is one that assists primarily through its social interactions, including verbal interactions and physical gestures. NAO is a 57-cm-tall humanoid robot, with four limbs, 25 degrees of freedom of movement, loudspeakers, microphones, a video camera, infra-red sensors, LEDs, force sensor units, inertial unit, sonars, joint position sensors, and contact and tactile sensors. NAO is being developed for pediatric rehabilitation (Carrillo et al., 2018).

Development has occurred over six years at the Royal Children's Hospital, Melbourne, Australia, with a team of rehabilitation consumers, health professionals, and engineers from Swinburne University of Technology. Development has included programming of exercises commonly prescribed by physical therapists, assessment of robot interactions with children and their families, and testing of feasibility and acceptability. A tablet-based interface for the selection of exercises, and a custom-built trolley allow deployment of NAO on the inpatient ward. The proposed uses target support of home- or ward-based exercise programs, augmentation of motivation, repetition, demonstration, and provision of immediate feedback and encouragement.

REHABILITATION EVALUATION

Pediatric populations being evaluated include individuals with cerebral palsy (CP), acquired brain injury, spinal cord injury, and oncological conditions. The robots are thought to be useful for periods of intensive rehabilitation, when there are sets of commonly prescribed exercises to be practiced multiple times daily.

Descriptive analysis regarding the robot experience indicates a positive response regarding attitude, perceived usefulness, likeability, and emotional reaction, as well as ease of use, enjoyment, sociability, and influence. There was a more varied level of acceptability as rated by physiotherapists, who had higher levels of anxiety and more concern regarding the robot's ability to support the rehabilitation session (Kruse et al., 2018).

Thematic analysis regarding impressions of the robot itself highlighted child enjoyment and positive affect, and parents' descriptions of utility for motivation, teaching, and potential for increased independent exercise. There was a wish for accessibility for use on weekends and at home. The users recognized a limitation related to a lack of active monitoring of performance. Participants still wanted human interaction and users noted the fact that the robot cannot engage in conversation (Butchart et al., 2021).

F-WORD LENS

Qualitative data have provided insights into social robot development for rehabilitation. Cutting-edge work with NAO, and ongoing related work in the treatment of autism and mental health issues (Pennisi et al., 2016), provide at times controversial fodder for thought. When applying the F-word lens, increased independence with exercise represents the potential for improved function. The fun factor is reflected in documented positive affective influence. Families are happy with the accessibility of therapy and expressed familiarity with the technology.

Fitness is promoted with delivery and teaching of repetitive exercises and there are reports of utility for increased motivation. Although friends are not directly assessed in this work, there were positive scores related to sociability, and one envisions the positive response of friends when people speak about their robot exercise assistant. The prospect of raising self-esteem with these sought-after devices deserves further thought.

Overall, the users express hope for having such devices as part of their future (Butchart et al., 2021). Families appreciate their child's life-long needs and struggle with limited service. There is hope for improvement in artificial intelligence (AI) capability to make robots more responsive to individual needs, and hope for access to social robots in the home and clinical settings to increase the intensity of therapy in the context of a busy family life. Similar to VR therapeutic games, families see a future with less time spent in advocacy for services or waiting on waiting lists. They see a future with therapist-directed intense, individualized, goal-focused therapy ready and waiting when required.

To reach the future as seen through the F-words lens, there is an indication for early and meaningful incorporation of users' wishes and a vision of utility as development moves forward, including enhancement of NAO's exercise skill, video monitoring of sessions, visual kinematic monitoring, real-time feedback on exercises, voice recognition, and the development of AI to improve interaction.

Along with these positive messages, the research data also send a clear message regarding a preference for human interaction. There is also variable acceptance by therapists. Careful quantitative and qualitative assessment is required, moving forward, as we ask: How can we positively incorporate robots into therapy programs? Can new-found independence be established? Does this technology help over-burdened therapists? Can it assist families in arduous care and tasks, freeing time for more positive interaction with family and friends?

The concept of "technology with heart," facilitating positive human interaction rather than replacing it, allows one to be excited to see how the future of social robots unfolds.

ROBOTIC WALKERS: THE SOLOWALK

THE TECHNOLOGY

This walker was designed by Carleton University's Advanced Biomechanotronics and Locomotion Lab in Ottawa, Ontario, Canada, in consultation with medical and rehabilitation communities. It has a built-in automated lift to facilitate transfer, a body-weight harness support system, powered omnidirectional wheels that augment effort, as well as remote controls to activate key functions of the device and customize speed and limits according to the patient's needs.

REHABILITATION EVALUATION

The study of this device has involved three phases. Phase I: A Case Study, described use by a 17-year-old male with bilateral spastic CP (BSCP), quadriplegia, GMFCS IV. Due to growth and an increase in weight and contractures, this youth had not walked in two years. The team, including the youth, his family, clinicians and engineers, was impressed when he easily transferred into the device and walked 40 meters during his first 60-minute session (McCormick et al., 2016)!

In Phase II, a qualitative pilot study included four individuals with CP, GMFCS IV, ages 14–22 years, and their families, healthcare professionals, and engineers. Thematic analysis documented physical and social advantages and disadvantages, with more advantages than disadvantages identified using the robotic walker. The youth liked the ability to walk with hands-free, and the potential to improve fitness. They dreamed of using the device for independence in the bathroom and for participation in sports. They asked for increased numbers of harness designs, smaller walker sizes for use in the home, and modifications to enable device use on uneven terrains (Alazem et al., 2020).

The third phase moved into the community. During fitness activities using the SoloWalk©, his heart rate moved into expected ranges for moderate to intense exercise, and qualitative data from youth and his family indicated positive opinions with regard to safety, comfort, and responsiveness of the device (McCormick et al., 2018). Feedback regarding possible challenges included the size of the device, which appeared large for homes and schools, and a price point that would currently affect equal accessibility.

TREXO PLUS©

THE TECHNOLOGY

The Trexo Plus© has wearable lower limb robotic devices with motors providing power for precise repetitive movement at the hips and knees. It utilizes a Rifton© walker base and can fit children from 2 to 12 years of age weighing up to 68 kilograms. The motion of the robotic device is controlled via a wireless tablet. The range of motion and gait speed are programmed individually according to each child's ability. The amount of support for weight bearing can be varied depending on the child's needs through the use of a saddle-style seat and chest prompt. The tablet records step count, speed, and progress toward mobility goals. Unlike previous large stationary pediatric exoskeleton-like devices, this device allows environmental discovery, with the ability to be used indoors as well as outdoors on dry flat surfaces such as sidewalks, pathways, and parks.

REHABILITATION EVALUATION

A one-time training session with the Trexo Plus© device was possible, involving four children aged two to three years of age who used a wheelchair as their primary mode of mobility with diagnoses including cerebral palsy, Angelman's syndrome, and arthrogryposis (McCormick et al., 2019a,b). There was subsequent interest in studying the device because therapist feedback was quite positive. The therapists involved were pleased with the ease of fit and adjustment, the ability of the child to explore their environment, the expressed feelings of security and safety in the walker, and the children's responsiveness to the robotic assistance. The therapists also noted the ability of the children to assist initiation of movement, which was smooth, gradual, precise, and repetitive. There was hope for the development of motor memory for reciprocal stepping with optimal alignment. It was also felt that this device offered a new exercise option for children with a high level of motor impairment, as children were fatigued after use. Therapists described the fact that gait training work within this user group would have been a challenge if it were not for the device, and that this device provided the possibility for intensive and frequent gait training therapy not available through traditional therapy programs. The registration and storage of data on the tablet, including settings for the range of motion at hips and knees, time spent in the device, speed, and step count were thought to be added benefits. It was felt that these data could be used for developing gait-related goals and measuring change over time with less therapist effort and time spent.

In addition, this session led to highlighted opportunities for improvement. For some children, there were challenges to positioning in the device while wearing orthosis. As well, the current device works for straight pathways, but requires hands-on assistance to turn and it does not stop when tone is increased. The therapist or assistant must therefore monitor the child when in the device. Note that study and development are ongoing at various research centers (Diot et al., 2023).

Further to these observations, a case study has illustrated the ability to increase step count in a 4-year old with CP GMFCS IV from zero to approximately 1,000 steps in a one-hour session. As well, the child's heart rate increased to moderate exercise intensity levels (McCormick et al., 2019a,b).

Research is currently in progress in the form of a randomized control, cross-over trial comparing the effect of Trexo© gait training and functional clinical therapy in children with cerebral palsy. To date, there is positive qualitative feedback from participants and their families. One parent describes their child as being much more motivated to use the device when outdoors, exploring in nice weather, particularly where there is a longer distance to travel on a straight path and a lower need for frequent assisted turns. Perceived gains reported by parents to date include improved posture and trunk control, as well as improved sleep patterns.

F-WORDS LENS

Robotic walkers may change the answer to a question commonly asked when families face a diagnosis of CP: "Will my child walk?" from "No" to "Yes, and this is how". Change in function has been documented as a first-time user of the SoloWalk© was able to transfer easily into the device and walk 40 meters after not walking for two years, while, in the Trexo Plus©, step count increased from zero steps to 1,000 steps in a one-hour session.

Qualitative data have demonstrated comfort and smiles on children's faces expressing the fun factor (McCormick et al., 2019a,b). A child with CP, GMFCS IV walked into school with their friends and found new ways of interacting with friends while playing basketball and soccer while using the Trexo© device. These fitness opportunities are combined with evidence of ability to progress from 10 meters of movement in the device to running a lap around an Olympic-sized gym. Heart rate elevation into moderate intensity exercise range is also early evidence of the potential impact on fitness for individuals with significant physical involvement (McCormick et al., 2018).

The voices of users were captured as they looked to their future with the hope that these devices will help them to manage private activities, such as going to the toilet independently, resulting in decreased strain on caregivers who still lift them, and enable increased access to fun activities, such as baseball in a standing position (Alazem et al., 2019).

The Call to Action

Let's drive this powerful vehicle of innovative engineering and technology forward within the field of rehabilitation and developmental medicine with specialized experienced teams that include the users, their families, clinicians, engineers, researchers, and business specialists. Let's shorten the time from idea conception to putting

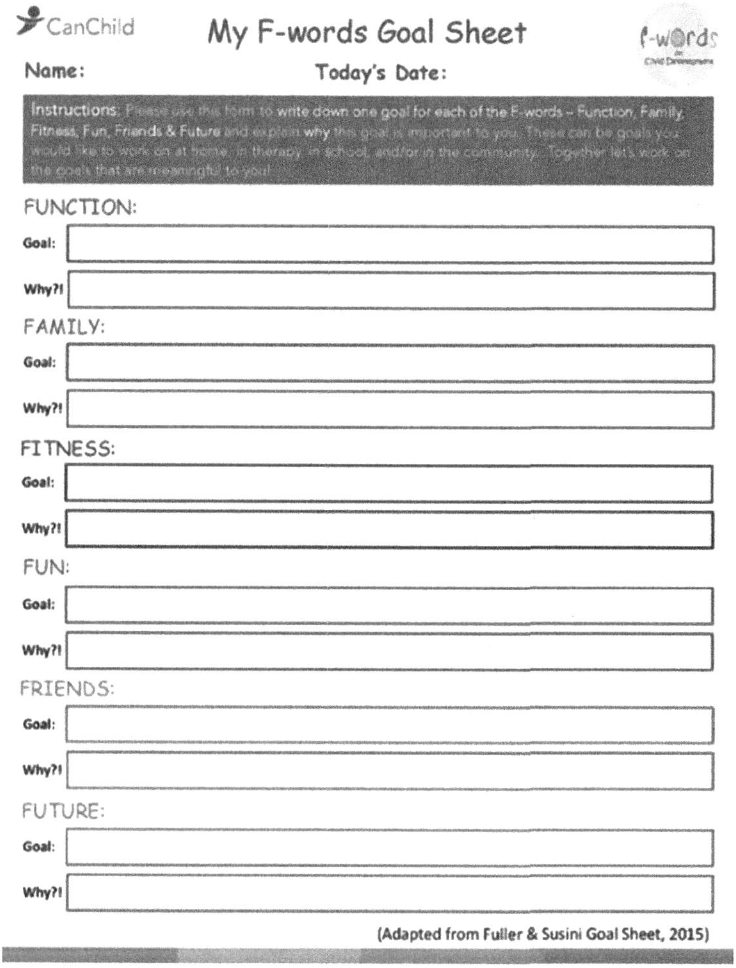

FIGURE 10.5   F-Words in Child Development Goal Sheet.

well-developed technology into the hands of the people who need it most. The first step in the development of technical solutions for children and individuals with disabilities is the identification of their needs, followed by the second step, which is the development of solutions by engineers (Mensah-Gourmel et al., 2022). Certainly, a lifetime to develop the perfect device is just too long! For ultimate success, it is essential to keep the child, youth and family at the center, building this technology using deep, holistic thinking. Careful reflection is required regarding goals that are meaningful to the users and their families (CanChild, 2020), as technology is designed to meet the complex needs of all individuals who face barriers that can be overcome. The "My F-Words Goal Sheet" can help identify goals, and articulate why these goals are important to children and their families (Figure 10.5).

Moving forward, it is critical that there is an optimistic vision refracted through the right lens. Let's change the conversation and use as many F-words as possible, because "I swear this is how we should think" (Rosenbaum and Gorter, 2012).

REFERENCES

Alazem, H., McCormick, A., Nicholls, S.G., Vilé, E., Adler, R., Tibi, G., 2019. Development of a robotic walker for individuals with cerebral palsy. *Disabil. Rehabilitation. Assist. Technol.*

Alazem, H., McCormick, A., Nicholls, S.G., Vilé, E., Adler, R., Tibi, G., 2020. Development of a robotic walker for individuals with cerebral palsy. *Disabil Rehabil Assist Technol* 15(6), 643–651. https://doi.org/10.1080/17483107.2019.1604827

Anaby, D., Avery, L., Gorter, J.W., Levin, M.F., Teplicky, R., Turner, L., Cormier, I., Hanes, J., 2020. Improving body functions through participation in community activities among young people with physical disabilities. *Dev Med Child Neurol* 62(5), 640–646. https://doi.org/10.1111/dmcn.14382

Biddiss, E., Chan-Viquez, D., Cheung, S.T., King, G., 2021. Engaging children with cerebral palsy in interactive computer play-based motor therapies: Theoretical perspectives. *Disabil Rehabil* 43(1), 133–147. https://doi.org/10.1080/09638288.2019.1613681

Biddiss, E., Knibbe, T.J., Fehlings, D., McKeever, P., McPherson, A., 2019. Positive distraction in pediatric healthcare waiting spaces: Sharing play not germs through Inclusive, hands-free interactive media. *Dev. Neurorehabilit* 22(7), 445–452.

Bonnechère, B., Jansen, B., Omelina, L., Jan, S.V.S., 2016. The use of commercial video games in rehabilitation: A systematic review. *Int J Rehabil Res* 39(4), 277–290. https://doi.org/10.1097/mrr.0000000000000190

Butchart, J., Harrison, R., Ritchie, J., Martí, F., McCarthy, C., Knight, S., Scheinberg, A., 2021. Child and parent perceptions of acceptability and therapeutic value of a socially assistive robot used during pediatric rehabilitation. *Disabil Rehabil* 43(2), 163–170. https://doi.org/10.1080/09638288.2019.1617357

CanChild, F. *Words Tools*. Available at: https://canchild.ca/en/research-in-practice/f-words -in-childhood-disability/f-words-tools (Accessed 20 June 2020)

Carrillo, F.M., Butchart, J., Knight, S., Scheinberg, A., Wise, L., Sterling, L., McCarthy, C., 2018. Adapting a general-purpose social robot for paediatric rehabilitation through in situ design. *ACM Trans Hum-Robot Interact Thri* 7, 12. https://doi.org/10.1145/3203304

Chen, Y., Fanchiang, H.D., Howard, A., 2018. Effectiveness of virtual reality in children with cerebral palsy: A systematic review and meta-analysis of randomized controlled trials. *Phys Ther* 98(1), 63–77. https://doi.org/10.1093/ptj/pzx107

Diot, C.M., Thomas, R.L., Raess, L., Wrightson, J.G., Condliffe, E.G., 2023. Robotic lower extremity exoskeleton use in a non-ambulatory child with cerebral palsy: A case study. *Disabil Rehabil Assist Technol 18*(5), 497–501.

Gong, D., Ma, W., Gong, J., He, H., Dong, L., Zhang, D., Li, J., Luo, C., Yao, D., 2017. Action video game experience related to altered large-scale white matter networks. *Neural Plast* 2017, 7543686. https://doi.org/10.1155/2017/7543686

Handler, L., Tennant, E.M., Faulkner, G., Latimer-Cheung, A.E., 2019. Perceptions of inclusivity: The Canadian 24-hour movement guidelines for children and youth. *Adapt Phys Act Q* 36(1), 1–18. https://doi.org/10.1123/apaq.2017-0190

Knights, S., Graham, N., Switzer, L., Hernandez, H., Ye, Z., Findlay, B., Xie, W.Y., Wright, V., Fehlings, D., 2016. An innovative cycling exergame to promote cardiovascular fitness in youth with cerebral palsy. *Dev Neurorehabil* 19(2), 135–140. https://doi.org/10.3109/17518423.2014.923056

Kruse, N., Butchart, J., Marti-Carillo, F., McCarthy, C., Rodda, J., Wise, L., Terling, L., Schienberg, A., 2018. *Acceptability of the NAO Humanoid Robot as a Therapeutic Aid in Rehabilitation, 9th Australasian Academy of Cerebral Palsy and Developmental Medicine.* Auckland, New Zealand, March.

Lewis, G.N., Woods, C., Rosie, J.A., Mcpherson, K.M., 2011. Virtual reality games for rehabilitation of people with stroke: Perspectives from the users. *Disabil Rehabil Assist Technol* 6(5), 453–463. https://doi.org/10.3109/17483107.2011.574310

McCormick, A., Alazem, H., Morbi, A., Beranek, R., Adler, R., Tibi, G., Vilé, E., 2016. Power walker helps a child with cerebral palsy. *Proceedings of the 3rd International Conference on Control, Dynamic Systems, and Robotics (CDSR'16)*. https://doi.org/10.11159/cdsr16.129

McCormick, A., Alazem, H., Vike, E., Freguson, K., Glossop, E., 2018. Robotic walker tested for exercise and participation in youth with cerebral palsy GMFCS IV. *9th Australasian Academy of Cerebral Palsy and Developmental Medicine.* Auckland, New Zealand, March.

McCormick, A., Biddiss, E., Scheinberg, A., Rocon, E., Alazem, H., McEwan, A., 2019. *Advancements in Neurorehabilitation Technology for Function, Activity and Participation: Hype or Hope?* Anaheim: American Academy of Cerebral Palsy and Developmental Medicine (AACPDM), September.

McCormick, A., Gaebler, D., Biddis, E., Argall, B., Rocon, E., Alazem, H., 2019. *Translating Neurorehabilitation Technology Advancements into Clinical Practice: Hype or Hope?* Chicago: American Congress of Rehabilitation, November.

McCormick, A., Wright, V., Alazem, H., 2019. *Innovative Robotic Technology Promotes Neurorehabilitation: Past, Present and Future.* Paris: European Academy of Childhood Disability (EACD), May.

Mensah-Gourmel, J., Bourgain, M., Kandalaft, C., Chatelin, A., Tissier, O., Letellier, G., Gorter, J.W., Brochard, S., Pons, C., Benyahia, A., Bréchoire, I., Julia, P., Nguyen-Luong, T.-X., Mirlesse, P., Studenik, T., Loriot, S., Tesar, B., Menn, N.L., Gayon, T., Genot, A., Ropars, J., Newman, C.J., Paradis, J., Houx, L., Bailly, R., Venineaux, M., 2022. Starting from the needs: What are the appropriate sources to co-create innovative solutions for persons with disabilities? *Disabil Rehabil Assist Technol*, 1–10. https://doi.org/10.1080/17483107.2022.2114554

Murphy, M.P.A., 2020. COVID-19 and emergency elearning: Consequences of the securitization of higher education for post-pandemic pedagogy. *Contemp Secur Policy* 41(3), 492–505. https://doi.org/10.1080/13523260.2020.1761749

Nguyen, L., Cross, A., Rosenbaum, P., Gorter, J.W., 2019. Use of the international classification of functioning, disability and health to support goal-setting practices in pediatric rehabilitation: A rapid review of the literature. *Disabil Rehabil* 43(6), 884–894. https://doi.org/10.1080/09638288.2019.1643419

Pennisi, P., Tonacci, A., Tartarisco, G., Billeci, L., Ruta, L., Gangemi, S., Pioggia, G., 2016. Autism and social robotics: A systematic review. *Autism Res* 9(2), 165–183. https://doi .org/10.1002/aur.1527

Porras, D.C., Siemonsma, P., Inzelberg, R., Zeilig, G., Plotnik, M., 2018. Advantages of virtual reality in the rehabilitation of balance and gait: Systematic review. *Neurology* 90(22), 1017–1025. https://doi.org/10.1212/wnl.0000000000005603

Ravi, D.K., Kumar, N., Singhi, P., 2017. Effectiveness of virtual reality rehabilitation for children and adolescents with cerebral palsy: An updated evidence-based systematic review. *Physiotherapy* 103(3), 245–258. https://doi.org/10.1016/j.physio.2016.08.004

Reyes, F., Niedzwecki, C., Gaebler-Spira, D., 2020. Technological advancements in cerebral palsy rehabilitation. *Phys Med Rehabil Clin N Am* 31, 117–129. https://doi.org/10.1016 /j.pmr.2019.09.002

Rosenbaum, P., Gorter, J.W., 2012. The 'F-words' in childhood disability: I swear this is how we should think! *Child Care Heal Dev* 38(4), 457–463. https://doi.org/10.1111/j.1365 -2214.2011.01338.x

Sevick, M., Eklund, E., Mensch, A., Foreman, M., Standeven, J., Engsberg, J., 2016. Using free internet videogames in upper extremity motor training for children with cerebral palsy. *Behav Sci (Basel)* 6(2), 10. https://doi.org/10.3390/bs6020010

Shikako-Thomas, K., Shevell, M., Lach, L., Law, M., Schmitz, N., Poulin, C., Majnemer, A., group, the Q., 2015. Are you doing what you want to do? Leisure preferences of adolescents with cerebral palsy. *Dev Neurorehabil* 18(4), 234–240. https://doi.org/10.3109 /17518423.2013.794166

Simeonsson, R.J., Leonardi, M., Lollar, D., Bjorck-Akesson, E., Hollenweger, J., Martinuzzi, A., 2003. Applying the international classification of functioning, disability and health (ICF) to measure childhood disability. *Disabil Rehabil* 25(11–12), 602–610. https://doi .org/10.1080/0963828031000137117

Soper, A.K., Cross, A., Rosenbaum, P., Gorter, J.W., 2019. Exploring the international uptake of the 'F-words in childhood disability': A citation analysis. *Child Care Heal Dev* 45(4), 473–490. https://doi.org/10.1111/cch.12680

Soper, A.K., Cross, A., Rosenbaum, P., Gorter, J.W., 2020. Service providers' perspectives on using the 'F-words in childhood disability': An international survey. *Phys Occup Ther Pediatr* 40(5), 534–545. https://doi.org/10.1080/01942638.2020.1726551

Soper, A.K., Cross, A., Rosenbaum, P., Gorter, J.W., 2021. Knowledge translation strategies to support service providers' implementation of the 'F-words in childhood disability'. *Disabil Rehabil* 43(22), 3168–3174. https://doi.org/10.1080/09638288.2020.1729873

Stucki, G., 2009. Current state of the application of the ICF in physical medicine and rehabilitation. *Int J Rehabil Res* 32, S30.

Stucki, G., Cieza, A., Ewert, T., 2001. Application of the international classification of functioning, disability and health (ICF) in clinical practice. *Phys Med Rehabilitationsmedizin Kurortmedizin* 11, 231–232. https://doi.org/10.1055/s-2001-19075

Stucki, G., Kostanjsek, N., Ustün, B., Cieza, A., 2008. ICF-based classification and measurement of functioning. *Eur J Phys Rehabil Med* 44(3), 315–328.

11 Opportunities to Reduce Inequities through Tele-wheelchair Assessments

The Importance of Co-designing Services with indigenous Māori with Lived Experience of Disability

Pauline Boland, Bernadette Jones, Laura Desha, Beauche McGregor, and Fiona Graham

With a new dawn comes a sense of wellbeing and optimism.

Unknown author, *Māori Whakatauki* (proverb)

OVERVIEW: THE WRITING TEAM

To explore the use of telehealth in wheelchair assessment from an Indigenous Māori perspective, we drew on Kaupapa Māori principles (Pihama et al. 2002), whereby knowledge is created with, by, and for Māori. In doing so, our intention was to focus on the concerns and needs of Māori, respect tikanga (traditional protocols), Māori knowledge, and te reo Māori (the Māori language). Following Māori traditions of sharing whakapapa (ancestry – who you are, where you come from), we begin with an introduction of the authors, to illustrate the perspectives underpinning this work.

- Pauline Boland is an Irish woman and an occupational therapist based at the University of Limerick, Ireland. Pauline worked in Aotearoa/New Zealand for many years where she collaborated with the other authors on research in the disability field.
- Bernadette Jones affiliates to Ngā Wairiki, Ngāti Apa iwi (Māori tribes). She has lived experience of disability and supports a spouse who is a wheelchair user. Bernadette has a background in nursing and is currently a

Māori health and disability senior researcher with the University of Otago, Aotearoa/New Zealand.

- Laura Desha is of Mauritian Indian and Welsh heritage and is an occupational therapist based in Brisbane, Australia. Laura uses telehealth platforms to engage with international researchers and research participants from disability services, including building connections and partnering with Māori and non-Māori at the University of Otago, Aotearoa/New Zealand.
- Beauche McGregor is affiliated with Ngāti Kahungunu and Rangitane iwi (Māori tribes). She has lived experience of disability and is a wheelchair user. Beauche has recently completed her Bachelor of Social Work and contributes her disability expertise to facilitating the independent voices of disabled Māori. (See Appendix 1 for full Pepeha/Māori introduction.)
- Fiona Graham is a pakeha (non-Māori) New Zealander of Irish, Scottish, and French descent and is an occupational therapist and academic. Her clinical experience and research span many areas of disability, specializing in rehabilitation with children with neurodisability and telehealth. Fiona works at the University of Otago, Aotearoa/New Zealand.

BACKGROUND

Global inequities in health and disability outcomes for Indigenous, disabled people[1] (King et al. 2014) are equally evident in the provision of assistive technology (Jones et al. 2017; Graham et al. 2022). Māori, the Indigenous people of Aotearoa/New Zealand, make up approximately 16.7 percent of the population (Statistics NZ 2020). Despite major health and disability system reforms[2] aimed at improving services across the system, inequities in health outcomes for Māori remain apparent (Crengle et al. 2022), arising from practices inherently linked to colonization and Western models of health service provision which may be experienced as alienating and hostile by Māori (Graham and Masters-Awatere 2020). This situation is compounded for tāngata whaikaha Māori (Māori with lived experience of disability) where the intersectionality of discrimination, being both Māori and disabled, can lead to multidimensional impacts for this population (Ingham et al. 2022a). Our earlier research suggests that an important contributor to these inequities has been a lack of culturally responsive services, whereas key facilitators included explicit provision of whānau (extended family) support, emotional care, and health system navigation (Graham et al. 2022, 2021). Common barriers to accessing services include remote geography and cost of travel for services, factors which tele-delivery of wheelchair and seating assessment has the potential to address.

Appropriate wheelchair and seating assessment is required to ensure people with mobility challenges can be afforded the same rights in society as non-wheelchair users (Gallagher et al. 2022; ISWP 2022). Wheelchair and seating assessments are often highly complex interventions, involving advanced skills in anatomical measurement, prediction of future muscular-skeletal change, product knowledge, and attention to social factors (Rousseau-Harrison and Rochette 2013). Healthcare professionals, predominantly occupational therapists and physiotherapists, require additional training beyond their pre-registration training courses to competently assess and prescribe wheelchair solutions (Mathis and Gowran 2021). Given the degree of specialty of these assessments, the difficulties accessing such a service are unsurprising, particularly for rural and remote communities, leading to telehealth as a potential way to enable wider access to expertise and timely service for wheelchair prescription. Policy and infrastructure to support wheelchair provision varies globally, with a range of gaps in service provision evidenced (Gowran et al. 2021), though this service is essential if the goals of the United Nations Convention on the Rights of People with Disabilities (UN, 2006) to promote full and effective participation and inclusion in society (UN 2006; Gallagher et al. 2022) are to be realized. Telehealth has been proposed as one strategy to bridge this gap between need and service.

Consistent with international research on telehealth in general (Khoja et al. 2005; Doraiswamy et al. 2020), our research on the acceptability of telehealth for wheelchair assessment in Aotearoa/ New Zealand has been positive, including for Māori wheelchair users (Graham et al. 2022). Telehealth services vary in how they are designed, from high to low technology, and asynchronous web-based activities to synchronous live-streamed consultations. Most commonly, telehealth wheelchair assessment involves a specialist assessor located at a distance from a wheelchair user and an onsite assistant (e.g., a family member, paid carer, or local health professional). Using video conferencing software, the specialist assessor guides the wheelchair user or onsite assistant to position the wheelchair user, undertake measurements, and collect other information. Recently, Ott et al. (2022) attributed the success of tele-delivered wheelchair assessment to the quality of the relationship between the remote support person and specialist assessor. Older observational and quasi-experimental studies have concluded that telehealth was at least equivalent to in-person assessment in the recommendations reached (Schein et al. 2010a) and the satisfaction of wheelchair users (Schein et al. 2010b). Health professionals tended to have mixed perspectives on the risks and value of telehealth compared with wheelchair users (Graham et al. 2020), a pattern seen in many aspects of telehealth research (Ott et al. 2022).

In addition, our research has identified patterns in Māori wheelchair users' acceptance and use (Williams et al. 2015) of telehealth (Graham et al. 2020; Graham et al. 2021; Graham et al. 2022). We focused on the most complex of wheelchair assessments, acknowledging that the current (largely in-person) system set a context for participants' views. Firstly, wheelchair users expressed deep frustration in gaining access to wheelchair and seating assessment in the current system, citing waiting periods of up to two years. Additionally, many wheelchair users experienced in-person assessment as a disempowering process in which they felt at the mercy of the system, with little voice in the timeliness or process of the assessment. Māori wheelchair users particularly noted the importance of cultural safety during teleconsultation, whakawhanaungatanga (building trusted relationships) in the online context, and the potential for summative disadvantage for Māori (Graham et al. 2022). The expectations from all stakeholders were not that telehealth would alleviate all current issues with the wheelchair assessment system, but rather that telehealth could improve access to assessment (and possibly more timely funding) as travel time would be minimized for assessors. Telehealth would also enable greater flexibility with who could attend sessions online, such as whānau (extended family) who could otherwise not be present, an important incentive for Māori in particular.

Wheelchair assessors in our research also raised some concerns about potential for compromised accuracy of assessment, as well as, conversely, reduced capacity to build relationships online. Given that research to date (albeit quasi-experimental) suggests equivalent assessment outcomes between in-person and telehealth wheelchair assessment (Schein et al. 2010a), this perceived risk may be unfounded. However, a dominant perspective voiced by wheelchair users was that the desire to gain timely access to assessment far outweighed their concern that some inaccuracies in assessment may occur. Randomized trials are needed to achieve certainty of these outcomes, but so too is a greater understanding of the context and mechanisms of successful telehealth wheelchair assessment. The lost potential for access to specialist rehabilitation services for Māori, like wheelchair assessment through telehealth (Graham et al. 2021), exacerbates existing inequalities. Research evidence indicates that the moral imperative is to offer telehealth to everyone, and, should they decline, to then proceed with in-person assessment.

Telehealth literature has shown that the benefits of this technology have been numerous in overcoming barriers for marginalized populations, with increased access to specialists, decreased financial burden, and reduced travel times (Ware 2013; Jones et al. 2017). Some key components of successful telehealth interventions for Indigenous populations include the importance of culturally tailoring programs, use of familiar mobile devices, and a well-trained Indigenous health workforce (Williams et al. 2017; Dawson et al. 2020). There appears to be moderate evidence for uptake and improved health outcomes for Indigenous people using online-designed interventions based on findings from a recent systematic review (Reilly et al. 2020). Recommendations for telehealth with Indigenous peoples include culturally relevant tailoring of content and presentation format, alongside the inclusion of Indigenous people in the design of new services (Wright et al. 2021; Smylie et al. 2022). If "delivering best practice remotely" is a key factor in the successful implementation of telehealth-delivered wheelchair assessment (Ott 2022), then we argue that including an Indigenous perspective on the process of assessment must be central to this design.

It is imperative that tāngata whaikaha Māori, with experience of wheelchair assessments, are at the forefront of any telehealth service design. These disabled Māori service users are optimistic about the potential of telehealth to redress the current power imbalance in the interaction with wheelchair assessors without risking relationship building, which both wheelchair users and assessors value. We have, therefore, used an Indigenous Māori model of health in this next section, illustrated with examples of the lived experiences of tāngata whaikaha Māori, to draw attention to the depth and holistic cultural considerations relevant to current and future telehealth services for Indigenous wheelchair users.

PRIVILEGING THE VOICES OF TĀNGATA WHAIKAHA MĀORI

In this section, three Indigenous Māori wheelchair users give voice to their experiences of wheelchair assessments, and their perspectives on these processes. Manawa[3] (pseudonym) and Claire (actual name) were interviewed as part of our earlier qualitative enquiry into wheelchair assessment (Graham et al. 2022); Manawa, a male of 45–55 years of age, had used a wheelchair for over six years at the time of being interviewed and Claire, a female of 36–45 years of age, had over ten years' experience as a wheelchair user. Beauche, a 28-year-old Māori woman, co-author of this chapter, and a wheelchair user since she was three years of age, was interviewed in a series of pūrākau (traditional storytelling) (Ingham et al. 2022b) sessions for this book chapter (Pepeha/introductions in Māori are available for Claire and Beauche in the appendix to this chapter). Pūrākau has its roots in traditional Māori ways of sharing experiences and fits with Māori-centered research as a means of gathering and exploring Indigenous narratives. It also privileges the voices of tāngata whaikaha Māori, enabling conversations to address inequities such as those engendered by ableism, institutional racism, and the impacts of colonization (Ingham et al. 2022a).

The narrative excerpts from interviews and pūrākau are presented using the framework of the "Meihana model" (Pitama et al. 2014). This model presents a seafaring journey toward the ultimate goal of hauora (health and well-being), using the analogy of the voyage of a waka hauora, a type of double hulled canoe in which Māori ancestors migrated to Aotearoa/New Zealand in pre-colonial times (Pitama et al. 2007). Through this model, we highlight diverse cultural and societal influences for Māori, both positive and negative, and these are described visually in Figure 11.1 below.

FIGURE 11.1   The Meihana model (Pitama et al. 2007).

The two connected hulls of the waka represent a client linked to their whānau. It is the role of a culturally competent health practitioner to board the waka (canoe or vessel) in a supporting role, to enable whakatere (navigation) through the health system journey to well-being, in this case, through tele-wheelchair assessment. While ocean-bound, the waka can be propelled and impacted by ngā roma moana (ocean currents), representing diverse experiences of being connected to te ao Māori (the Māori worldview), as described below. Ocean currents may be harnessed to benefit the journey, whereas the waka is at risk of being blown off-course by the challenging, interrelated historical and current societal forces of the ngā hau e wha (the four winds, a metaphor indicating the negative impacts of colonization, marginalization, migration, and racism). For healthcare providers, the Meihana model challenges those delivering services to consider such influences upon clients. Culturally, Māori are a collective people, rather than individualistic, and the strong linkage between the client and whānau are shown by the crossbeams between the two hulls of the waka. These represent holistic elements of health and well-being, which are critical to clients and whānau in a comprehensive and culturally responsive health encounter. These interwoven elements include tinana (physical well-being), hinengaro (psychological and emotional well-being), wairua (beliefs regarding connectedness and spirituality), taiao (the physical environment), and ratonga hauora (access to services /systems that support health). Each component of the Meihana model is relevant to all healthcare contacts, whether delivered in-person or via telehealth.

NGĀ ROMA MOANA (OCEAN CURRENTS): WHĀNAU, WHENUA, AHUA, AND TIKANGA

The ocean currents describe elements of Māori identity, reflecting te ao Māori which may act as protective factors that support well-being and emphasize the interconnectedness of all things spiritual and physical, the creation stories of the natural world and tāngata whenua (people of the land), and tikanga.

Telehealth services, with embedded and tangible affirmations of te ao Māori offer a foundation to address both racism and ableism by providing culturally safe experiences for tāngata whaikaha Māori. The Meihana model notes four elements of identity that provide a protective effect on well-being and are key to building culturally successful client–clinician relationships.

Whānau are foundational to Māori cultural identity, and hence are integral to any culturally competent service for Māori. Among other benefits, whānau provide a support network and draw on their collective strength to protect and build resilience for individual members of their whānau. Relationships with clients and whānau have potential implications for every aspect of a client's engagement with telehealth, impacting priorities, values and beliefs, decision-making, and the roles that will be assumed in such an encounter (e.g., who will provide emotional or practical support).

Whenua is the genealogical or spiritual connection between a client (and/or their whānau) and land, and is fundamental to Māori identity, and hence their well-being.

Talking about the connection to land is one means of building strong connections and relationships and can be done readily via telehealth.

Ahua are described as signs of connectedness with a Māori identity. Ahua applies to both the client and health professional including use of te reo Māori (Māori language), and visible cultural symbols, such as the wearing of a taonga (cultural treasure), although cultural responsiveness is not contingent upon outward signs of ahua. Health professionals who have developed skills in cultural responsiveness, can show they are cognizant of ahua by respectfully enquiring about, and meeting, some of the cultural needs of Māori clients (Graham and Masters-Awatere 2020).

Considering the Meihana model, service users are bound closely to whānau, and are interconnected either through having common ancestry or through sharing a common purpose (Lawson-Te Aho 2010). Beauche, who has used a wheelchair since early childhood, described the support and advocacy role assumed by her whānau (family) through largely in-person engagement with health professionals:

> From birth to the age of 18, my whānau have supported me, especially when I was younger, they were able to articulate my needs with health professionals. Having whānau there sheltered me, without them it would have been harder, and I wouldn't have got the outcomes I got…

This support was critical for Beauche's voice to be heard in a non-Māori, complex multi-layered system in which services and staff appeared under extreme pressure.

> Even my mum found the wheelchair assessment process very frustrating, I wouldn't have been able to manage without her. My mum pushed and pushed for better seating for me but the amount of pushback we got from the assessors[4] was outrageous.

A culturally competent, well-resourced workforce would respond to client preferences for whānau involvement, and incorporate the needs, strengths, and aspirations of whānau. The practical efficiencies of telehealth for whānau are readily apparent, such as the ease of involving immediate and extended family members who are geographically dispersed or need flexible meeting times. Yet telehealth assessment may not be desirable for all, and offering choice about its timing and use is integral to whānau having agency in the journey to well-being (Graham et al. 2022).

Telehealth services may enable prominence of elements of te ao Māori providing, for example, overt messages of inclusion such as placement of a familiar Māori design on a video-call background. Clients' specific personal preferences for cultural support can also be efficiently met by accessing diverse and often geographically remote resources. As Manawa stated:

> [it] … does open the opportunities to have someone there who may not otherwise come to the visit like a Māori provider or an interpreter if you wanted to converse in te reo Māori. You could better coordinate that over the video link.

Tikanga describes Māori customs, including principles, values, and norms, which offer guidance for moral behavior (Durie 1998; Justice 2001). Health professionals working with Māori in Aotearoa are alerted to key aspects of tikanga as a way of ensuring the cultural safety of Māori clients, including the centrality of trusted relationships prior to launching into clinical tasks, such as wheelchair assessment. Whakawhānaungatanga extends beyond the familiar Western concept of health professionals building rapport with clients, and focuses on building mutually respected trusted relationships. Claire noted,

> At the beginning of this call [conducted by a Māori interviewer] … you stated where you were from and that was really cool, and then you gave me the opportunity to do that as well, and I feel I have an understanding of your history, of your whakapapa [ancestry]. That's really awesome …. So, you're not just a person on a screen; I know also about your history.

Although some service providers are aware of the basic practical elements of tikanga, lack of skill in more nuanced observation of customs undermines cultural safety, as Manawa described:

> I felt [the home assessment] breached tikanga in a way because … I mean, so I got them to take [their] shoes off at the door. But you know, tikanga is more than that … do you feel you are happy to have them in your house? And I can say that I was not happy to have them in the house. And just the whole way they sort of quite often stood over you or you know….

Beauche suggests that telehealth may address such a power imbalance:
"Using a zoom would give you more control than going into a hospital for assessment." Manawa elaborated:

> I think having it [wheelchair assessment] done by a video link could improve the safety of these consultations. You know you've got to protect [yourself]. You can just hang up if you feel unsafe or you have control in that moment, because it's a lot more even playing field than when therapists are in your house.

Tāngata whaikaha Māori using telehealth services, which not only accommodate but also embrace customary practices, and are sensitively responsive, may take up opportunities to observe personally relevant tikanga and translate these into the virtual realm. Beauche proposes, "We could ensure we use basic tikanga practices like opening and closing with a karakia [blessing or prayer]. Telehealth provides an opportunity for the user to set the cultural foundation more [than] going to their [hospital]."

NGĀ HAU E WHA (THE FOUR WINDS): COLONIZATION, MARGINALIZATION, MIGRATION, AND RACISM

The four winds are a maritime metaphor for describing four forces with historical and ongoing negative impacts on Indigenous people. Colonization impacted heavily on Māori, resulting in loss of most of their tribal lands and forcing the migration of

Māori to cities to seek housing, food, and employment. Post-colonization, Māori experienced marginalization from historically being a collection of different iwi (tribes) to that of a homogenized, minority population and they still endure negative health and social impacts arising from racism in current times (Graham and Masters-Awatere 2020). For tāngata whaikaha Māori, the negative effects of racism are amplified when poverty and being Māori co-occur (Ingham et al. 2022b). Manawa discussed implications for accessing telehealth infrastructure for some Māori:

> I'm very aware of the digital connectivity issue. The number of Māori who don't have access to technology, fast internet, or can't afford it, those kinds of things. So, I think that you're looking at summative disadvantage. Yeah. In terms of socio-economic disadvantage and structural discrimination and the disadvantage of disability there could be some that it [telehealth] just can't work for.

Tāngata whaikaha Māori, like other disabled people, have rights explicitly outlined by the UN (2006) in the Convention on the Rights of Persons with Disabilities, where the focus stresses respect for dignity and freedom to make ones' own choices. Yet, wheelchair and assessment systems have often lacked scope for client choice and control, particularly for Māori users where the compounding historical impacts of disadvantage and discrimination can be experienced. Beauche expressed a desire for wheelchair and seating assessment in her adolescence to have been less directive, and more collaborative, reflecting on how,

> [They] try and dictate what's best for the clients. I find it very frustrating, and I resist against it. I know in some circumstances they are right, but they need to think about, this is our day-to-day life and we deal with our life the best that we can, and we know how to. I just feel like I don't get heard … Surely there could have been a better way. Don't tell me, work with me.

Working together is contingent on respectful and trusting relationships, key to mitigating and eliminating the negative impacts of the four winds on the health and well-being of tāngata whaikaha Māori. Claire described why the development of trust in healthcare providers, be they online or in-person, is a long-term process:

> I feel very strongly about, regarding the sense of a disabled person and our needs being completely side-lined … you have to look at 25 years now of literally fighting every day for everything that we have and that we need … It's a fight and it's really hard. We're always going to be sceptical, we can be quite a difficult group to deal with because there's a lot of distrust and a lot of anger, especially with people who have been in the situation for a while now; we don't trust the process.

Transparency in service design is another means of overcoming some of the negative effects of ngā hau e wha, to address the understandable cynicism which Māori have engendered regarding motivations for a transition to telehealth services:

I would be concerned [about telehealth wheelchair assessment] ... I'd hate to see this going into effect and it just not working and it be another, just yet another thing that we have to contend with purely because of cost-cutting measures, and for me that's not good.

(Claire)

Claire also described the high stakes and emotional burden of change, such as a shift to tele-delivered services:

It's really complicated, and if you get it [the wheelchair assessment] wrong then we're screwed, because how do we get anywhere, how do we work? How do we run our families or whatever it is? So, there's actually quite a bit at stake here for us, and that could cause quite a bit of anxiety.

Yet there is excitement about the potential for telehealth to make improvements on current wheelchair and seating services. "I think I'm really excited about a novel service delivery model for this because it actually starts breaking down the 'this is the way it happens' paradigm." (Manawa)

A commitment to cultural safety across integrated services and sectors is important, as Manawa states: "There is a big disconnect in te ao whaikaha [the worldview of disabled Māori] between health and disability services. Disability services don't understand te ao Māori in general." Health and disability services have a responsibility to address inequities by putting choice and control at the fore of the design of tele-delivered wheelchair assessment. To prioritize the Indigenous user voice, ratonga hauora need to include mechanisms that value and facilitate these voices. This was not the experience of Beauche where she commented, "I've never had a feedback check after any of my wheelchair assessments. I've never been asked about my cultural needs, how the process went, or what needs improving."

Wheelchair assessments should not be burdensome and disruptive to users but rather they should be flexible and fit in with people's lives, as Beauche clearly stated when discussing the impact of time spent in clinics being assessed, "Actually, I have a life. I have places to be, people to see, things to do."

DISCUSSION AND RECOMMENDATIONS

POTENTIAL OPPORTUNITIES TO ADDRESS INEQUITIES FOR MĀORI AND INDIGENOUS PEOPLE

The opportunities of telehealth for wheelchair and seating assessment for Māori wheelchair users have been presented, framing these voices within the Meihana Model, and informed by our earlier research (Graham et al. 2022), highlighting that tele-delivered assessment for wheelchair users could contribute to overcoming summative disadvantage experienced by tāngata whaikaha Māori. There is considerable dissatisfaction with existing wheelchair services, including assessment and provision of equipment, related not only to lack of access to specialists in this field, but also to healthcare professionals who are culturally responsive in both in-person and online healthcare interactions (Graham et al. 2022). Tele-delivered specialist services, such as complex wheelchair assessments, have considerable potential to improve access and equity of services, not only for Māori but also other Indigenous communities, but only if delivered in the context of culturally safe and competent practice.

Like Chaet et al. (2017), our research identified that, for telehealth interactions to achieve their potential, people must have access to appropriate resources (technology and internet availability) as well as healthcare professionals who are skilled and confident in addressing the health issue, and in using telehealth technology (Graham et al. 2022). We support these ideas as the basis for all Indigenous wheelchair users, with comfort during assessment often brought about by specific, planned, culturally sensitive techniques and attitudes embedded within the telehealth interactions.

The right-to-choice (in this case, the choice of telehealth-delivered wheelchair assessment as standard) is a key recommendation within the United Nations Convention for the Rights of People with Disabilities (UN, 2006). Given the relative novelty of tele-health-delivered wheelchair assessment, many people with disabilities are not aware of this as an option (Valdez et al. 2021). Our first recommendation is that health services observe the right of all people, particularly Indigenous peoples, to routinely offer well-resourced and culturally safe telehealth-delivered wheelchair assessments. While the evidence base in this field continues to evolve, the current evidence for its effectiveness presented thus far, alongside wider literature about digital design with Indigenous groups, offer helpful templates for service design which are potentially applicable to wheelchair assessment (Dawson et al. 2020; Reilly et al. 2020).

OPPORTUNITIES FOR CO-DESIGNING CULTURALLY APPROPRIATE TELE-WHEELCHAIR SERVICES

Our second recommendation is to ensure full-service user involvement in the design of current and future telehealth services for wheelchair users, specifically using methodologies consistent with Indigenous worldview and acknowledging the historical influences on all healthcare interactions (Povey et al. 2021), to avoid replication of existing services which are arguably "acultural," with related reinforcement of barriers for Indigenous wheelchair users. Indigenous healthcare workers are an important group to involve in designing future training (Violette et al. 2021) and supporting service users to transform the delivery of these services. For non-Indigenous health practitioners, an essential first step would be to develop understanding and acceptance of the continued impacts of colonization and related racist structures entrenching disparities (Smylie et al. 2022). In Aotearoa/New Zealand, ongoing involvement of Māori in the design and implementation of telehealth-delivered wheelchair services, to ensure these are 'mana-enhancing' (respectful and strength-based) rather than deficit-focused, will be vital to their success. One core way of promoting cultural safety for Māori and non-Māori staff and service users of such services is to embed Indigenous models of healthcare within such services, such as the Meihana Model. Despite the diversity of Indigenous cultures, these principles have international relevance.

CULTURAL SAFETY TRAINING

Our third recommendation is that training of healthcare professionals providing telehealth-delivered wheelchair assessment needs a broader scope, extending far beyond the specific technological skills needed to enact remote assessment (Martel et al. 2020; Reilly et al. 2020). We advocate for training in culturally responsive practice to be an integral, fundamental part of the development of tele-delivered wheelchair assessment services, with specific attention to risks and opportunities arising through the use of telehealth platforms. To address the challenges facing Indigenous people, this training must tackle the complexity and tensions inherent to a realization that "Cultural safety is not a technical skill, but a deep understanding of unconscious bias and inequities of access to care ... (Martel et al. 2020, p. 277). Both service users and health professionals in our research have indicated preferences for very simple, software-as-service technology that people already know how to use, thus minimizing the need for advanced technology skills. Yet, focusing only on the technical software and hardware elements required (Kim et al. 2021) will miss the opportunity for transformative advances in culturally inclusive healthcare to genuinely reduce inequities for Indigenous peoples (Graham et al. 2022). While international sharing of training resources offers a means of highlighting solutions, diversity in the challenges faced by Indigenous peoples will necessitate local resource development. The ongoing work by the International Society of Wheelchair Professionals (ISWP 2022), on designing and delivering training for assessors, could include links to training in individual countries for working with Indigenous wheelchair users. The authors are presently developing a training module in culturally responsive wheelchair assessment for Aotearoa/New Zealand.

STRENGTHS AND LIMITATIONS

One of the strengths of using the Meihana model as a foundation for this chapter is that we have used this framework to analyse and interpret the voices of Indigenous, marginalized people who are often not visible in a Westernized system of health. Additionally, the inclusion of Māori people, with lived experience of disability and a wheelchair user, as both authors and participants in our studies added to the validity of the interpretations and enabled amplification of often marginalized voices. Although we drew on the experiences of only three such individuals directly in this chapter, and the conclusions must be considered in relation to the contexts the narratives were gathered in, we argue that the key points we present have the potential to be applied to other Indigenous people using wheelchair services.

CONCLUSION

This chapter illustrates that contemporary research and practice philosophies, congruent with Indigenous methodologies and worldviews, are essential to truly maximize telehealth services and break down existing inequities in healthcare service design (Chaet et al. 2017). Reflection on culturally responsive telehealth service design is important to achieve equitable health outcomes. In an era of the burgeoning use of telehealth, an opportunity exists to optimistically embrace a "new dawn:" Telehealth can be designed to optimize choice, reduce inequities, enhance flexibility, and progress rights-based services for Indigenous wheelchair users.

APPENDIX 1

With permission, the following participants' pepeha are presented below. A pepeha is the traditional way that Māori people introduce themselves and it gives reference to the geographical landmarks that they connect to, including mountains, rivers, oceans, and lakes. A pepeha also allows the individual to share their genealogy and pay homage to their ancestors. One participant, Manawa, chose to remain anonymous.

BEAUCHE MCGREGOR

Ko Rangitumau toku maunga
Ko Ruamāhanga toku awa
Ko Tākitimu toku waka
Ko Hurunui o Rangi me Te Ore Ore oku marae
Ko Ngati Kaiparuparu me Ngati Hāmua oku hapu
Ko Ngati Kahungunu ki Wairarapa me Rangitane o Wairarapa oku iwi
No whakaoriori au
Ko Beauche McGregor ahau

CLAIRE FREEMAN

No reira tēna koutou tena koutou tēna tatou katoa.
Ko Ngātokimatawhaorua tōku Waka
Ko Ngāpuhi tōku iwi
Ko Manaia tōku Maunga
Ko Hatea tōku Awa
Ko Ngāti korokoro Raua ko Ngāti poa Ngā hāpu
Ko Kokohuia tōku Marae
i te taha o tōku pāpā
Ko Kiriora te Waenga ōku Tipuna kua noho ki Hokianga
Ko Claire Freeman tōku ingoa

NOTES

1. The term "disabled people" has been adopted by the New Zealand health system. In this chapter, we use the "Social Model of Disability" in recognition that people are disabled by society (Lorelli 2004).
2. In a reform process in 2022, the New Zealand Ministry of Health has been restructured, resulting in the replacement of District Health Boards with a more coordinated, culturally responsive health system, "Te Whatu Ora – Health New Zealand," and "Te Aka Whai Ora – Māori Health Authority," and the creation of a new ministry, "Whaikaha – Ministry of Disabled People."

3. Manawa preferred to remain anonymous for his contribution to this chapter, whereas Claire chose to reveal her identity.
4. A term used to describe healthcare professionals who assess for wheelchair and seating equipment, most typically occupational therapists or physiotherapists.

REFERENCES

Chaet, D., Clearfield, R., Sabin, J. E. and Skimming, K. (2017) 'Ethical practice in telehealth and telemedicine', *Journal of General Internal Medicine*, 32(10), 1136–1140.

Crengle, S., Davie, G., Whitehead, J., de Graaf, B., Lawrenson, R. and Nixon, G. (2022) 'Mortality outcomes and inequities experienced by rural Māori in Aotearoa New Zealand', *The Lancet Regional Health-Western Pacific*, 28, 100570.

Dawson, A. Z., Walker, R. J., Campbell, J. A., Davidson, T. M. and Egede, L. E. (2020) 'Telehealth and indigenous populations around the world: A systematic review on current modalities for physical and mental health', *Mhealth*, 6, 1–17.

Doraiswamy, S., Abraham, A., Mamtani, R. and Cheema, S. (2020) 'Use of telehealth during the COVID-19 pandemic: Scoping review', *Journal of Medical Internet Research*, 22(12), e24087.

Durie, M. (1998) *Te Mana, Te Kawanatanga: Politics of Māori Self Determination*, Auckland: Oxford University Press.

Gallagher, A., Cleary, G., Clifford, A., McKee, J., O'Farrell, K. and Gowran, R. J. (2022) '"Unknown world of wheelchairs" A mixed methods study exploring experiences of wheelchair and seating assistive technology provision for people with spinal cord injury in an Irish context', *Disability and Rehabilitation*, 44(10), 1946–1958.

Gowran, R. J., Bray, N., Goldberg, M., Rushton, P., Barhouche Abou Saab, M., Constantine, D., Ghosh, R. and Pearlman, J. (2021) 'Understanding the global challenges to accessing appropriate wheelchairs: Position paper', *International Journal of Environmental Research and Public Health*, 18(7), 3338.

Graham, F., Boland, P., Grainger, R. and Wallace, S. (2020) 'Telehealth delivery of remote assessment of wheelchair and seating needs for adults and children: A scoping review', *Disability and Rehabilitation*, 42(24), 3538–3548.

Graham, F., Boland, P., Jones, B., Wallace, S., Taylor, W., Desha, L., Maggo, J., McKerchar, C. and Grainger, R. (2022) 'Stakeholder perspectives of the sociotechnical requirements of a telehealth wheelchair assessment service in Aotearoa/New Zealand: A qualitative analysis', *Australian Occupational Therapy Journal*, 69(3), 279–289.

Graham, F., Boland, P., Wallace, S., Taylor, W. J., Jones, B., Maggo, J. and Grainger, R. (2021) 'Social and technical readiness for a telehealth assessment service for adults with complex wheelchair and seating needs: A national survey of stakeholders', *New Zealand Journal of Physiotherapy*, 49(1), 31–39.

Graham, R. and Masters-Awatere, B. (2020) 'Experiences of Māori of Aotearoa New Zealand's public health system: A systematic review of two decades of published qualitative research', *Australian and New Zealand Journal of Public Health*, 44(3), 193–200.

Ingham, T., Jones, B., Perry, M., King, P. T., Baker, G., Hickey, H., Pouwhare, R. and Nikora, L. W. (2022a) 'The multidimensional impacts of inequities for Tāngata Whaikaha Māori (Indigenous Māori with lived experience of disability) in Aotearoa, New Zealand', *International Journal of Environmental Research and Public Health*, 19(20), 13558.

Ingham, T., Jones, B., Toko King, P., Smiler, K., Tuteao, H., Baker, G. and Hickey, H. (2022b) 'Decolonising disability: Indigenous Māori perspectives of disability research in the modern era', in Rioux, M. H., Viera, J., Buettgen, A. and Zubrow, E., eds., *Handbook of Disability*, Singapore: Springer.

ISWP (2022) *International Society of Wheelchair Professionals*, https://wheelchairnetwork.org/

Jones, L., Jacklin, K. and O'Connell, M. E. (2017) 'Development and use of health-related technologies in indigenous communities: Critical review', *Journal of Medical Internet Research*, 19(7), e7520.

Khoja, S., Casebeer, A. and Young, S. (2005) 'Role of telehealth in seating clinics: A case study of learners' perspectives', *Journal of Telemedicine and Telecare*, 11(3), 146–149.

Kim, H.-S., Kim, H. J. and Juon, H.-S. (2021) 'Racial/ethnic disparities in patient-provider communication and the role of e-health use', *Journal of Health Communication*, 26(3), 194–203.

King, J., Brough, M. and Knox, M. (2014) 'Negotiating disability and colonisation: The lived experience of Indigenous Australians with a disability', *Disability & Society*, 29(5), 738–750.

Lawson-Te Aho, K. (2010) *Definitions of Whanau: A Review of Selected Literature. A Families Commission Report*, Welllington, https://thehub.swa.govt.nz/assets/documents/definitions-of-whanau_FC%20_10.pdf

Martel, R., Reihana-Tait, H., Lawrence, A., Shepherd, M., Wihongi, T. and Goodyear-Smith, F. (2020) 'Reaching out to reduce health inequities for Māori youth', *International Nursing Review*, 67(2), 275–281.

Mathis, K. and Gowran, R. J. (2021) 'A cross-sectional survey investigating wheelchair skills training in Ireland', *Disability and Rehabilitation: Assistive Technology*, 1–8, Early Online.

Ministy of Justice (2001) *He Hinātore ki te Ao Maori; A Glimpse into the Maori World*, Wellington: Māori Perspectives on Justice, https://www.justice.govt.nz/assets/Documents/Publications/he-hinatora-ki-te-ao-maori.pdf

Ott, K. K., Schein, R. M., Straatmann, J., Schmeler, M. R. and Dicianno, B. E. (2022) 'Development of a home-based telerehabilitation service delivery protocol for wheelchair seating and mobility within the Veterans Health Administration', *Military Medicine*, 187(5–6), e718–e725.

Pihama, L., Cram, F. and Walker, S. (2002) 'Creating methodological space: A literature review of Kaupapa Maori research', *Canadian Journal of Native Education*, 26(1), 30–43.

Pitama, S., Huria, T. and Lacey, C. (2014) 'Improving Māori health through clinical assessment: Waikare o te Waka o Meihana', *The New Zealand Medical Journal*, 127(1393), 107–119.

Pitama, S., Robertson, P., Cram, F., Gillies, M., Huria, T. and Dallas-Katoa, W. (2007) 'Meihana model: A clinical assessment framework', *New Zealand Journal of Psychology*, 36(3), 118–125.

Povey, J., Raphiphatthana, B., Torok, M., Nagel, T., Shand, F., Sweet, M., Lowell, A., Mills, P. P. J. R. and Dingwall, K. (2021) 'Involvement of indigenous young people in the design and evaluation of digital mental health interventions: A scoping review protocol', *Systematic Reviews*, 10(1), 1–8.

Reilly, R., Stephens, J., Micklem, J., Tufanaru, C., Harfield, S., Fisher, I., Pearson, O. and Ward, J. (2020) 'Use and uptake of web-based therapeutic interventions amongst Indigenous populations in Australia, New Zealand, the United States of America and Canada: A scoping review', *Systematic Reviews*, 9(1), 1–17.

Rousseau-Harrison, K. and Rochette, A. (2013) 'Impacts of wheelchair acquisition on children from a person-occupation-environment interactional perspective', *Disability and Rehabilitation: Assistive Technology*, 8(1), 1–10.

Schein, R. M., Schmeler, M. R., Holm, M. B., Saptono, A. and Brienza, D. M. (2010a) 'Telerehabilitation wheeled mobility and seating assessments compared with in-person', *Archives of Physical Medicine and Rehabilitation*, 91(6), 874–878.

Schein, R. M., Schmeler, M. R., Saptono, A. and Brienza, D. (2010b) '*Patient* satisfaction with telerehabilitation assessments for wheeled mobility and seating', *Assistive Technology*, 22(4), 215–222.

Smylie, J., Harris, R., Paine, S.-J., Velásquez, I. A. and Lovett, R. (2022) 'Beyond shame, sorrow, and apologies—Action to address indigenous health inequities', *British Medical Journal*, 378, o1688.

Statistics NZ (2020) *2018 Census Totals by Topic – National Highlights (updated)*, Wellington, https://www.stats.govt.nz/information-releases/2018-census-totals-by-topic-national -highlights-updated.

UN (2006) *Convention on the Rights of Persons with Disabilities*, https://www.un.org/development/desa/disabilities/convention-on-the-rights-of-persons-with-disabilities.html

Valdez, R. S., Rogers, C. C., Claypool, H., Trieshmann, L., Frye, O., Wellbeloved-Stone, C. and Kushalnagar, P. (2021) 'Ensuring full participation of people with disabilities in an era of telehealth', *Journal of the American Medical Informatics Association*, 28(2), 389–392.

Violette, R., Spinks, J., Kelly, F. and Wheeler, A. (2021) 'Role of indigenous health workers in the delivery of comprehensive primary health care in Canada, Australia, and New Zealand: A scoping review protocol', *JBI Evidence Synthesis*, 19(11), 3174–3182.

Ware, V. (2013) *Improving the Accessibility of Health Services in Urban and Regional Settings for Indigenous People*, Canberra: Australian Institute of Health and Welfare.

Williams, M. D., Rana, N. P. and Dwivedi, Y. K. (2015) 'The Unified Theory of Acceptance and Use of Technology (UTAUT): A literature review', *Journal of Enterprise Information Management*, 28(3), 443–488.

Williams, M. H., Cairns, S., Simmons, D. and Rush, E. (2017) 'Face-to-face versus telephone delivery of the green prescription for Maori and New Zealand Europeans with type-2 diabetes mellitus: Influence on participation and health outcomes', *New Zealand Medical Journal*, 130, 71–79.

Wright, A. L., VanEvery, R. and Miller, V. (2021) 'Indigenous mothers' use of web- and app-based information sources to support healthy parenting and infant health in Canada: Interpretive description', *JMIR Pediatrics and Parenting*, 4(2), e16145, https://doi.org /10.2196/16145.

12 Person-centered Perspective on the Use of Technology in Healthcare

Sanne Angel

INTRODUCTION

THE USE OF TECHNOLOGY AS A CORE HUMAN COMPETENCE

There is no doubt that the ability to use tools is essential to *Homo sapiens* and its domination of the Earth. The use of technology is a core human competence that has enabled the human world to move far away from nature, for better or worse. From the proverb "Necessity is the mother of invention," we can derive the essence of how technology has worked to facilitate human life and expression. In combination with the fascination of human nature with being able to create, technological development occurs in a flow, where what appears to be needed is mixed with what turns out to be possible. This means that only later does it become clear whether technology promotes or hinders human life. The development of technology in its own right thus creates possibilities that are not necessarily good – merely possible. This is not necessarily an assessment that is even contemplated. Over the past decade, technological developments have been explosive and, as usual, in the midst of all the enthusiasm about what it is possible to develop, we may have lost sight of how it affects the world of human beings. This equally applies to technological solutions in the healthcare service. Should an assessment be made, a crucial factor would be whether the value of the assessment was done in terms of saving resources or qualifying efforts. In any case, it is time for these ethical assessments to be systematically undertaken, so that both citizens and professionals can know what they are being exposed to and exposing others to. In focusing on the person-centered perspective, it may be helpful to divide the analysis into the intended function of the technology, what it cannot do, and any unintended effects.

To shed light on what technology means from the perspective of the person, we will draw upon the phenomenological tradition, with its particular strength in gaining insight into the world in the way it makes sense to people, also called the lifeworld. The German philosopher Martin Heidegger, in particular, has contributed insights into what it means to be human, for which reason this philosophical approach will be central in the unfolding of how the meaning that people attach to

DOI: 10.1201/9781003272786-12

technology depends on the situation: what you need help for, and whether it is a question of survival or of regaining lost abilities that enable you to function.

TECHNOLOGY IN HEALTHCARE

Knowledge of illness is part of the education of all healthcare professionals, and at best they can also draw on their own experiences, provided these are well processed. But as the human beings that professionals also are, attention to their own lives and the demands of their work may mean they do not have the time to see things from the patient's perspective. It may also escape the professional's attention that the patient is often debilitated and thus not in his or her habitual state, and this position is reinforced by the patient's lack of scientific knowledge and preconceptions about health compared with the professionals', which hampers their ability to perceive and interpret events. In cases where the illness or injury threatens the patient's identity, the stakes for the patient are overwhelming, which also affects their ability to see clearly and to act (Ashworth et al. 1992). On top of this comes the patient's limited knowledge of the practice in which the patient's influence must occur. Such knowledge is hindered precisely because the patient lacks full insight into the group's practice and knowledge, knowledge of the group's members, and ownership of the group's language. The patient is in an unfamiliar context, which also makes it difficult for professionals to confirm the patient's original identity. The framework is thus neither conducive to the patient being recognized as a person, nor is it conducive to the patient being able to recognize the professionals as people. This makes it more difficult to be treated as an individual which, in turn, makes it difficult to adapt professional standards so that they promote quality for the individual rather than conformity due to standardization (Ashworth et al. 1992). This distance from the human being in the patient is exacerbated by the production logic of our time and the optimization ideal of managing many tasks with scarce resources: better, faster and cheaper is the goal, and here technology offers itself as a means. And while it may well be the means, it still needs to be investigated if it is in line with the ideal of professional aids.

THE IDEAL OF HEALTHCARE PROFESSIONALS

The healthcare professional ideal of securing a good life for everyone dates back to ancient Greece, as we can read from the writings of the physician Hippocrates. This ideal is still upheld, as is underlined by the reminders of the UN (2022) to nations and professionals: no one must be left behind. However, there is a clear difference in the answers to the question of what the good life is, which implies that only the individual can find his or her way to a meaningful life. This understanding is central to Heidegger's (1962) philosophy and emerges in recent definitions of rehabilitation (WHO 2021; Wade 2020; Maribo et al. 2022) and basic nursing theories (Henderson 1995), in which the meaningful life is presented as the goal. Thus, beyond the concrete task, the healthcare professional service is aimed at the life that can be lived when the healthcare professionals have completed their task and the patient can once again take control. This is true even if this control is not complete, but must be supported by other people, whether relatives or paid helpers. It follows that achieving a meaningful life for the patient implies that the patient must be granted control to the greatest extent possible. This places special demands on the professional caregiver to intervene on behalf of the patient only for as long as it is needed and to step back as soon as it is possible (Heidegger 1962 §26). Being someone who supports the patient's existence, without trying to take it over, is the only way to ensure that the intervention promotes a meaningful life for the patient. Only the patient can decide whether the direction is the right one to fulfill his or her potential, especially when illness or injury rules out previous possibilities for the future and new ones must be found (Angel et al. 2009a, 2009b, 2011). Being in control is necessary in order to get to the point where one's potential can be fulfilled (Heidegger 1962). This underlines the importance of the emphasis on patient participation in recent decades (Angel & Frederiksen 2015). At the same time, however, other people play a major role in the way forward towards self-understanding and being understood in life, as well as in the possibilities that might challenge you in finding your new potential.

TECHNOLOGICAL GAINS IN RELATION TO ILLNESS

Technological developments in healthcare have contributed to a reduction in morbidity and mortality, and a consequent increase in life expectancy. Technology in healthcare embraces knowledge, skills, medication, instruments, and machines, although artificial innovations are what most people understand by the term. Technology can play a very special role in treating illness and injury, both for survival and in the long term. When illness or injury strikes, the patient is confronted with a present reality that was not on in the cards yesterday and is not a natural extension of the past just before the event. Likewise, the future, as it appeared to offer itself, is gone. In the hope of regaining this, technology is important in many ways; if you have a spinal cord injury, you hope for technological developments that can reduce the damage; if you have cancer, you hope for the invention of a new cure. It is thus well known that people in serious situations tend to hope for technological developments when no such solutions are realistic at the present time. Technology and digitalization are in vogue at present, due to an extremely promising development that the world could not previously even have imagined. In the same way, technology is explored in an attempt to solve the imbalance between the numerous people that face a large number of problems and the inadequate human and financial resources available to solve them.

Lifesaving Technological Developments

Lifesaving technical innovations are the core of the intensive care unit (ICU) and can make the difference between life and death (e.g., respirators, cardiac defibrillators) and enable people to stay alive. Accordingly, patients and relatives willingly submit to what has been called 'the dictatorship of technology' due to the difficulty of mastering technology that requires extensive training and demands total focus from nurses (Wilkin & Slevin 2004; Pereira et al. 2014; de la Fuente-Martos et al. 2018; Alasad 2002). This is accepted by patients, relatives, and nurses alike, because they acknowledge the lifesaving function of the technology. In addition, technological skills are an essential part of intensive care nursing (Wilkin & Slevin 2004), making nurses feel secure and in control, and are valued for supporting the patient (Alasad 2002). The technology is so important that it is accepted that it can lead to patients being seen as "things" (Alasad 2002), putting lifesaving technology before humanity, at least in the emergency situation, and thereby becoming a barrier to humanism. But lifesaving technology may also ensure survival and a meaningful life in the long term.

TECHNOLOGY AS LIBERATION AND FREEDOM

The development of long-term technological solutions to supplement the human body and reduce the demands of everyday life picked up pace after the First World War. The exclusion of the injured from participation was remedied by enabling mobility, which could be regained by means of medical aids, such as the wheelchair, and through prosthetics such as artificial legs, which in some cultures determined whether a woman could be offered marriage because she could perform the expected tasks of a housewife. Another example is the possibility through technology of ameliorating the isolation brought about by loss of speech, by supporting communication. Technology can thus extend a person's abilities where the physical and mental impact involves loss of function: impairment and thus disability. Heidegger (1962) describes the optimal degree of integration with a tool as being like a hammer, which can become one with the person who uses it. In this case, the dividing line between the hammer and the hand is no longer clearly conscious. The positive gains are capabilities that would otherwise be lost. Thus, for some people, the technology can become so integrated with their bodily existence that the technology becomes them.

TECHNOLOGY AS STIGMATIZING

Technology as a positive gain in capabilities may come at a cost, however, due to the loss of the beauty of the pure body and of normality in general. To take the example of a toddler who needs glasses, my own daughter hated them, threw them out of the window, and was only happy when, as an adolescent, she could get contact lenses. The adaptation process might have gone better if her kindergarten teacher, like a friend of mine and her colleagues, had created a whole demonstration in which the teachers put on glasses, perhaps just in their hair, and got the children to draw, cut out and use brightly colored cardboard glasses. This experience of not wishing to appear unattractive or stand out negatively also emerged from my study of rehabilitation after spinal cord injury, where it took a 60-plus-year-old woman more than a year to harness the liberating function of a mobility scooter. She could not bear the thought of appearing in the town as a disabled person, as her grandson put into words when he asked with interest if she was going to get a "spasser scooter." (I hardly need add that a "spasser" is a derogatory term for a person with spasticity.) The fear of not being able to present oneself and be recognized as a human being with a disability was also the reason why a woman in her late 20s left the support bandages she had been given to facilitate the function of her spastic hands in the car when she went to meet work contacts. These are examples of how the gains of optimized functioning can be overshadowed by the limitation of the possibilities of "who I can be." The degree of inclusion in the cultural context is crucial both for how the person with a disability is encountered, and for how it is possible to think about oneself and find a meaningful way of being a person with a disability. At the same time, technology plays a positive role in leveling the inequality between the able-bodied and the disabled. The gap between wishes and possibilities thus provides a direct link to the search for a technological solution, not least in the struggle to conquer the limitations of illness and disability. However, the use of technology may demand professional support (Craddock & McCormack 2002)

TECHNOLOGY AS REPLACEMENT FOR THE OTHER

We will now take a closer look at what it is that technology cannot replace – the particular needs that can only be met by what is particularly human. What continues to make a difference between a human being and a machine is that being human gives the possibility of empathizing with the other human's situation (Stendevad 2016). This capacity for empathy has been the subject of extensive philosophical studies (Husserl 1960; Zahavi 2014) and involves a sensitive openness to the being of the other, which allows the other to understand himself or herself on the basis of the resonance that this calls forth in the helper. This occurs in the context of the presence of the helper as a caregiver. It is thus not only the understanding of the other that is important for the patient's self-understanding, but also the compassion and the willingness to be the strong one who takes on the task, so that the patient in their weakened condition can rely on the helper, in the confidence that they will be cared for and have their needs met.

TECHNOLOGY AS SUBSTITUTE FOR PHYSICAL HELP

In addition to more and more tasks being automated, machines have been developed to replace the help that has traditionally been provided as a care task by a caregiver to a person in need. This has made it possible to train muscles in a person paralyzed by spinal cord injury that cannot be trained through exercise. At the same time, the possibility of training the body without the use of human helpers means that there are no limits on the optimal training. However, staff happen not to use the machines (Howard et al. 2022). In another example, a bathing robot was not welcomed by nursing staff (Beedholm et al. 2015). The reason may be found in that personal hygiene assistance is also called personal care; the use of the word "personal" implies that this has a particular orientation toward the individual. This places value on the special qualities of the individual's needs and, at the same time, appears to include an assumption that this particularity can only be addressed by the particularity of the human being, which is what constitutes the difference between a human and a machine.

Care is so closely associated with solicitude that, when this is expressed in two words, they are often joined by an "and:" "care and solicitude." This can be read with Heidegger's distinction between the task being undertaken (*Besorge*, solicitude) and the care provided by another (*Fürsorge*, care) (Heidegger 1962 §26). This implies that the act of caring for someone can be performed without solicitude for the other. The central issue here may be that the helpful factor lies in the compassion (Angel 2021).

With the bathing robot taking over the care task, the concern of the nurses and myself was that, in their weakened state, patients unable to take care of their own personal hygiene would be left without solicitude. On reflection, in my otherwise well-functioning life, I personally would have preferred to be able to help myself by using the bathing robot. This would free me from having to relate to another person and organize my schedule accordingly. So, in theory, the person in need should welcome this opportunity for greater independence, liberation, and freedom. The risk of this assessment is that we once again overlook what is at stake for the weakened person, as described by Ashworth et al. (1992): for does a weakened person not need solicitude? In a study of young men with muscular dystrophy, it was clear that all they wanted was their caregiver's practical help in managing bodily functions (Martinsen & Dreyer 2012). The solicitude they received came from their social networks. On the other hand, they were unfortunately so weakened that they would not be able to operate the bathing robot themselves, and thus the use of this technology made almost no sense. This means that, in the investigation of the bathing robot, the analysis needs to divide into the bathing and being treated in concordance with one's need as a person.

With this discussion, I have illustrated the importance of careful consideration of the citizen's needs and whether the use of the technology makes sense in the individual situation and actually contributes to a liberating experience of independence, rather than to the person being left to their own devices, in the worst-case scenario. This includes the extension that the other can contribute both physically and mentally, and the significance of the relationship.

TECHNOLOGY AS SUPPORT FOR INTERSUBJECTIVITY

The development of electronically transmitted images from the district nurse's consultation with the hospital surgeon (Clemensen 2004, 2005) has led to a large increase in online-only consultations. These may be combined with physical consultation as seen in hearth rehabilitation (Knudsen et al. 2019, 2020). This has been particularly used where large geographical distances limit the possibility of physical attendance, as is, for example, the case in Australia. With the necessary shift away from physical meetings due to the COVID-19 epidemic, the use and adoption of such technology accelerated, so that the need for physical meetings became reconsidered in terms of the time, effort, and cost savings in relation to the need for transport and possible accommodation away from home (see, for example, the difference between teaching a class in the same physical room or online in Angel & Dahl 2020)

It is not entirely clear what the difference is between a physical meeting and an online meeting when it comes to interaction. Seeing a psychologist online, for example, can be experienced as rewarding (Hoffmann et al. 2018), and may feel less intimidating without the direct eye contact and physical presence of an unfamiliar physical setting. For a professional like me, however, this seems completely back to front, as direct eye contact and physical presence are considered to be the basic elements of empathy in human interaction: the eye that not only sees and provides the basis for overview and holistic understanding but which also recognizes, provides reassurance, love, courage, and encouragement. Physical presence is what creates a basis for sensing how the other is feeling and provides safety, relaxation, affirmation, and reassurance that together we can do it.

I have personal experience of perceiving a student's loss of self-confidence at a screen meeting, and of providing the necessary support. But I had seen the student's expression in her eyes and face previously in real life, and I was thus already familiar with the student's problem. At the same time, we had built up a mutual trust, so she did not turn away, but remained open to my help. But would the same thing have happened at an online meeting if a nurse colleague had sensed that the partner of a heart surgery patient was withdrawing from the interaction? What if the partner had simply withdrawn from the screen, vulnerable and unable to cope with the difficult situation or receive help for it, and had thus been beyond the reach of the nurse's support, so that the nurse had not observed it at all? In this way, observing through a screen limits the conditions for the nurse to follow the patient and maintain the necessary contact, as well as the nurse's possibility to undertake assessment, because the nurse receives a limited view of the patient in the form of the face and part of the upper body, and the background chosen by the patient. It is thus not a total picture of the patient with his or her body, interacting with the environment with motor skills and attention. This may be a larger loss than we imagine for the importance of reading the patient's situation using all senses for the nursing (as well as other healthcare professionals) assessment.

So, what do we know about interaction through a screen? One study shows that, in an online meeting with a group of people, there is significantly less eye contact than there would otherwise be (Tanaka 2022), and there is thereby a measurable difference. My own observation of online meetings is that one risks becoming a spectator rather than a participant, with the meeting chairperson attempting almost in vain to elicit comments and input. Here, then, is one difference that can be measured if it is tested. If we adopt an exploratory attitude to direct patient–nurse/healthcare professional contact, then two people are sitting in separate locations without physical contact with each other. They have planned the meeting and, as with a physical meeting, they have prepared for it and may have an idea of what will come out of it. Prior to the meeting they have prepared themselves and made themselves presentable, at least with regard to what can be seen on the screen. They have set up the equipment and ensured that the internet connection was working and have made sure to keep an eye on the time. Now, they sit synchronously, but not physically, opposite each other, in front of a screen that shows the head and shoulders of the other. They can hear each other's voices and engage in dialogue, alternately speaking and listening. Even though we know that our gaze should be directed towards the little green light of the camera in order for the other person to look directly into our eyes, we look at each other's picture and into each other's eyes in the picture. The picture on the screen makes the meeting participant feel very self-conscious: do I look all right? is my face showing the right expression, or do I look bad? This self-awareness provokes self-assessment and at times a performance – unlike at a meeting, where one's whole attention is directed at the other: who are they? what are they saying? This full attention implies self-transposition, where it is all about the other as described by the nurse philosopher Kari Martinsen (2003). The other person discloses their state of mind to the listener that thereby can resonate with their emotions and actually be felt. This applies primarily to the physical meeting, but can also be established in the telephone meeting, where the sound of the other person's voice creates the contact (Angel 2013). This gives meetings without screens a special strength, in terms of being there for the other. Concluding, technology is a "means" like tool to enable intersubjectivity without physical presence, but we need to be aware that it is not a physical presence.

TECHNOLOGY AS SELF-AID

Despite the challenges, technological solutions primarily represent a positive contribution that is here to stay, especially in a time when there is so much that needs to be done to improve human life, but relatively few resources. Free online resources for self-help can be provided to parts of the world, populations, or individuals that would otherwise have no assistance, such as videos on training in bladder emptying after a spinal cord injury (MossRehab 2022). In addition, the technology has enabled new innovative approaches that have not previously been offered.

But even though most people are adept at using online facilities, it is still a jungle to find the right one quickly. This also applies to those patients who possess the necessary knowledge. It is important not to underestimate the challenges of understanding one's own body, psyche, existence, and situation as seen in a study of people recovering from stroke (Kitzmüller et al. 2019). In the case of my own sprained ankle, the internet and my general professional understanding of the injury did not help as, after a month, the pain had only grown worse. It was only when a doctor performed a physical examination and held a dialogue with me about how to understand the pain and react to it that I was ready to order my daily life and my expectations toward the future: how I could move around, which of my daily activities and sports I could resume, and at what level. Would I be able to resume dance training and, if so, when, and could I still look forward to hiking?

Unfortunately, there seems to be little room for this empathetic presence in the quest of the health and social care system for efficiency, in the sense of rapid task completion. Whether this is ultimately cost-effective is something I would venture to question, as it entails the risk of imposing superficial and short-term solutions. The limited time available and the technological possibilities have led to a wide range of support systems in relation to decision-making and recovery processes (Gynea 2020; Hoffmann et al. 2018; Howard et al. 2022; Knitza et al. 2022; Knudsen et al. 2019; MossRehab 2022). It is documented that users of these systems are subject to selection bias, which means that participants are often the well-functioning patients. Despite these positive results, we must still ask ourselves whether self-help can replace the other person, or even the expert, interested conversation partner? If we look more closely at the importance of the other in the significance of narrative for a person's understanding of themselves and their situation, we know that dialogue is crucial in allowing the intervention to be adapted to the individual. As narratives enable access to the patient´s perspective, they allow the healthcare professional to develop an intervention that is tailored to the patient and which supports the patient in incorporating this into his or her narrative, providing a platform for future action. The patient's benefit from the intervention will materialize only if the patient understands and accepts the healthcare professional's messages, and if the solutions offered are tailored to the patient's situation. Then the patient knows what to do, considers this reasonable, and is able to do it (Angel 2021). According to Ricoeur (1983), individuals understand themselves through the narratives they build, and this builds on the task of hermeneutics and entails interpretation until a meaningful narrative is configured. Such a narrative enables an understanding of the self and the situation and paves the way for future action. In Ricoeur's (1983) theory of mimesis I, II and II, mimesis I is life as it is experienced. This implies an understanding of the world and the life of actions prior to the narrative, and the narrative is founded on this reality. The narrative configuration, mimesis II, is relating these facts into a narrative through a combination of signs and sentences. In the process of configuring this narrative in a way that makes sense, the other contributes by listening and questioning until he or she can follow the narrative (Ricoeur 1983). Mimesis III is the re-figuration of the life lived with actions that are founded on the narrative understanding gained from the configuration of mimesis II. This means that the narrative approach is helpful in eliciting the patient's story, which should be accomplished before relating to the problem as experienced by the patient, and finally ensures that the suggestion for action fits into the patient's perspective and is doable and manageable. This contrasts with the situation when the patient encounters standardized, internet-based interventions. In this case, the patient is left alone with his or her interpretation without having anyone to present it to, which implies the risk of misinterpretation and self-deception. The patient may initially feel helped, but later discover that the underlying problem remains unaffected. Some internet-based initiatives are therefore supplemented with personal contact. This applies, for example, to a six-week free learning and coping program for women after treatment for ovary cancer (Gynea 2020). We do not know if this help is sufficient, but, compared with receiving no help at all, it is appreciated.

In offers involving technology as a replacement for the other, it is also necessary to take into account the challenges presented by the disease, for which there may not be sufficient resources and insight to manage. Of particular note here is the assumption that the patient is his or her habitual self, which, as Ashworth et al. (1992) point out, is risky. This is not even to mention the patients who, even in their best moments, would not be able to use such an initiative and would thus be excluded.

INTEGRATING TECHNOLOGY IN CARING INTERTWINED WITH CONSIDERATION OF HUMAN EXISTENCE

Technology is, at one and the same time, alienating and empowering, challenging and reassuring for both patients and healthcare professionals. The constant evolution of technology requires the focus of healthcare professionals in order to use it, which takes their attention away from the patient. However, Alasad (2002) found that when ICU nurses' technological competencies increased, their attention shifted toward the patient and the patient's family, while the equipment faded into the background. Consequently, nurses need to master the technology in order to transcend it and thus be able to focus on the patient as a human being (Wallis 2005). This is especially important because a nurse can minimize the alienating effect of the technology on the patient by maintaining a close and supportive presence and providing comfort and care (Stayt et al. 2015). The nurse can do this by using human touch, being present, and using the patient's name as key features (McCallum 2002). The challenge of competence in dealing with the technical equipment makes sense in the ICU, but this may be one explanation for why purchased technology may still go unused (Howard et al. 2022). There can also be differences in technical skill, probably reinforced by generational divides, in which some professionals find the technology exciting and straightforward, whereas others cannot really get to grips with it and worry that valued practices are being lost.

Depending on the nature of the service, the technology should be analyzed in terms of its intended function, what it cannot do, and its unintended impacts. That said, there is no doubt that technology can be a great supplement, as long as the intention is not to phase out the patients' necessary contact with the healthcare professional as the expert and interested conversation partner. This assessment is based on a professional assessment that takes into account the fact that the patient may lack insight into what is going to help best. Once again, we face the debilitating and unknown situation that comes with illness.

Accordingly, a technological solution will work best for those patients who are able to meet the technology's requirements and have a high level of intellectual functioning, despite their illness. They must be able to read, understand, and follow the instructions, and assess whether the solution is applicable to them personally. Still, voice-based social support is perceived as stronger that text-based (Liu et al. 2022). As most patients are not equipped for this self-assessment, a professional will be needed. It is precisely this combination of technology and personal care that could optimize the benefits of technological solutions. My favorite scenario is that an elderly, debilitated citizen could meet with a consistent, personal professional in the citizen's own home or in municipal premises close by, and here, via a video link, meet with the therapists who are responsible for their separate areas, e.g., heart disease and cancer, and who, at the same time, can reach a common agreement. This would be a major benefit, rather than the patient having to go to several departments and perhaps different hospitals and keep track of information and plans that may even be contradictory.

Unfortunately, the inequality in the world means that those who perhaps are most in need possess neither the necessary knowledge, equipment, access, nor resources. Integration of the new technology thus becomes another source of increased inequality. It thus suggests the need to remind healthcare professionals to take due care and precautions. At the same time, it is, however, important to remember that the problem of inequality cannot be solved in the healthcare system. It is probably not possible to mitigate the worst consequences in a healthcare system that is subject to the same optimization processes that we know from industry. In addition, it is important to understand that inequalities between people's life situations, and thereby their health, are conditioned by cultural and societal factors, which in turn are reinforced by economic policies that support wealth and health for the richest.

CONCLUSION

In this chapter, the use of technology is acknowledged as a human core competence. It is acknowledged that technology is the source of development and wealth in human lives whereas technological progress also bears the risk of destruction, maybe even of the earth. The paradox is also known within health technology as a source for better and longer life as well as a risk toward the very core of humanity. Thus, the benefit of the use of technology as liberation and freedom as well as technology as a replacement of the other calls for integrating technology in caring intertwined with considering human existence. This leaves the professionals with the responsibility of assessing the technological solutions, the intended function of the technology, what it cannot do, and the unintended effect – never ever forgetting that the capabilities of humans may neglect human capability.

REFERENCES

Alasad J (2002) Managing technology in the intensive care unit: The nurses' experience. *Journal of International of Nursing Studies*, 39(4): 407–413. doi:10.1016/s0020-7489(01)00041-4

Angel S (2013) Grasping the experience of the other from an interview. Self-transposition in use. *International Journal of Qualitative Studies on Health and Well-Being*, 8(9): 1–7. doi:10.3402/qhw.v8i0.20634

Angel S (2021) Helpful factors in a healthcare professional intervention for low-back pain -unveiled by Heidegger's philosophy. *Nursing Philosophy*, 23(1): e12364. doi:10.1111/nup.12364

Angel S & Dahl MR (2020) Synchronous Distance Learning (SDL); an integrated course design using video conferencing for student-centered learning. https://health.medarbejdere.au.dk/#news-17583

Angel S & Frederiksen KN (2015) Challenges in achieving patient participation: A review of how patient participation is addressed in empirical studies. *Journal of International of Nursing Studies*, 52(9): 1525–1538. doi:10.1016/j.ijnurstu.2015.04.008

Angel S, Kirkevold M & Pedersen BD (2009a) Rehabilitation as a fight. A narrative case study of the first year after a spinal cord injury. *International Journal of Qualitative Studies on Health and Well-Being*, 4(1): 28–38. doi:10.1080/17482620802393724

Angel S, Kirkevold M & Pedersen BD (2009b) Getting on with life following a spinal cord injury: Regaining meaning through six phases. *International Journal of Qualitative Studies on Health and Well-Being*, 4(1): 39–50. doi:10.1080/17482620802393492

Angel S, Kirkevold M & Pedersen BD (2011) Rehabilitation after spinal cord injury and the influence of the professionals' support (or lack thereof). *Journal of Clinical Nursing*, 20(11–12): 1713–1722. doi:10.1111/j.1365-2702.2010.03396.x

Ashworth PD, Longmate MA & Morrison P (1992) Patient participation: Its meaning and significance in the context of caring. *Journal of Advanced Nursing*, 17(12): 1430–1439. doi:10.1111/j.1365-2648.1992.tb02814.x

Beedholm K, Frederiksen K, Frederiksen AM & Lomborg K (2015) Attitudes to a robot bathtub in Danish elder care: A hermeneutic interview study. *Nursing and Health Sciences*, 17(3): 280–286. doi:10.1111/nhs.12184

Clemensen J, Larsen SB & Bardram J (2004) Developing pervasive e-Heatth for moving experts from hospital to home. *IADIS International Journal on WWW/Internet*, 2(2). https://www.iadisportal.org/ijwi/vol2_2.html

Clemensen J, Larsen SB & Ejskjær N (2005) Telemedical treatment at home of diabetic foot ulcers. *Journal of Telemedicine and Telecare*, 11(Supplement 2). https://journals.sagepub.com/doi/abs/10.1258/135763305775124830

Craddock G & McCormack L (2002) Delivering an AT service a client focused social and participatory service delivery model in assistive technology in Ireland. *Disability and Rehabilitation*, 24(1–3). doi:10.1080/09638280110063869

de la Fuente-Martos C, Rojas-Amezcua M, Gómez-Espejo MR, Lara-Aguayo P, Morán-Fernandez E & Aguilar-Alonso E (2018) Humanization in healthcare arises from the need for a holistic approach to illness. *Medicina Intensiva*, 42(2): 99–109. doi:10.1016/j.medin.2017.08.002

Gynea (2020) *Norway*. Attended 2022-12-16. https://gynea.no

Heidegger M (1962) *Being and Time*. Harper & Row, New York

Henderson VA (1995) *The Nature of Nursing, Overvejelser Efter 25 år*. (The nature of nursing, reflections after 25 years). Munksgaard, Copenhagen

Hoffmann D, Rask CU, Hedman-Lagerlöf E, Ljótsson B & Frostholm L (2018) Development and feasibility testing of Internet-delivered acceptance and commitment therapy for severe health anxiety: Pilot study. *JMIR Mental Health*, 5(2): e28–e28. doi:10.2196/mental.9198

Howard J, Fisher Z, Kemp AH, Lindsay S, Tasker LH & Tree JJ (2022) Exploring the barriers to using assistive technology for individuals with chronic conditions: A meta-synthesis review. *Disability and Rehabilitation: Assistive Technology*, 17(4): 390–408. doi:10.1080/17483107.2020.1788181

Husserl E (1960) *Cartesian Meditations*. Trans. D. Cairns. Martinus Nijhoff, The Hague

Kitzmüller G, Mangset M, Evju AS, Angel S, Aadal L, Martinsen R, Bronken BA, Kvigne K, Bragstad LK, Hjelle EG, Sveen U & Kirkevold M (2019) Finding the way forward – The lived experience of people with stroke after participation in a complex psychosocial intervention. *Qualitative Health Research*, March: 1–14. doi:10.1177/1049732319833366

Knitza J, Muehlensiepen F, Ignatyev Y, Fuchs F, Mohn J, Simon D, Kleyer A, Fagni F, Boeltz S, Morf H, Bergmann C, Labinsky H, Vorbrüggen W, Ramming A, Distler JHW, Bartz-Bazzanella P, Vuillerme N, Schett G, Welcker M & Hueber AJ (2022) Patient's perception of digital symptom assessment technologies in rheumatology: Results from a multicentre study. *Frontiers in Public Health*, 22(10): 844669. doi:10.3389/fpubh.2022.844669

Knudsen MV, Laustsen S, Petersen AK, Hjortdal VE & Angel S (2019) Experience of cardiac telerehabilitation: Analysis of patient narrative. *Disability and Rehabilitation*, 12: 1–8. doi:10.1080/09638288.2019.1625450

Knudsen MV, Petersen AK, Angel S, Hjortdal VE, Maindal HT & Laustsen S (2020) Tele-rehabilitation and hospital-based cardiac rehabilitation are comparable in increasing patient activation and health literacy: A pilot study. *European Journal of Cardiovascular Nursing*, 19(5): 376–385. doi:10.1177/1474515119885325

Liu J, He J, He S, Li C, Yu C & Li Q (2022) Patients' self-disclosure positively influences the establishment of patients' trust in physicians: An empirical study of computer-mediated communication in an online health community. *Frontiers in Public Health*, 25(10): 823692. doi:10.3389/fpubh.2022.823692

Maribo T, Ibsen C, Thuesen J, Nielsen CV, Johansen JS & Vind AB (2022) *Whitebook on Rehabilitation (Hvidbog Om Rehabilitering)*. Rehabiliteringsforum Danmark, Aarhus

Martinsen B & Dreyer P (2012) Dependence on care experienced by people living with Duchenne muscular dystrophy and spinal cord injury. *Journal of Neuroscience Nursing*, 44(2): 82–90. doi:10.1097/JNN.0b013e3182477a62

Martinsen K (2003) *Fra Marx Til Løgstrup: Om Etikk og Sanselighet I Sykepleien*. From Marx to Loegstrup: On ethics and sensuousness in nursing. Universitetsforlaget, Oslo

McCallum C (2002) Balancing technology with the art of caring. *New Zealand Nurses 1995-, Kai Tiaki*, 8(7): 21–23

MossRehab (2022) *Philadelphia*. Attended 2022-12-12. https://www.mossrehab.com/scivideos

Pereira MMM, Germano RM & CÃ¢mara AG (2014) Aspects of nursing care in the intensive care unit. *Journal of Nursing UFPE*, 8: 545–554. doi:10.5205/reuol.5149-42141-1-SM.0 803201408

Ricoeur P (1983) *Time and Narrative I.* The University of Chicago, Chicago

Stayt LC, Seers K & Tutton E (2015) Patients' experiences of technology and care in adult intensive care. *Journal of Advanced Nursing*, 71(9): 2051–2061. doi:10.1111/jan.12664

Stendevad K (2016) *Fremtiden er Feminin - De 7 Spilleregler i det 21. Århundredes Lederskab (The Future Is Feminin – the Seven Rules in Leadership of the 21thcentury)*. Gyldendahl Business, Copenhagen

Tanaka S (2022) *Tokai University I Tokyo*. Presentation at the 39th International Human Science Research Conference 2022. Building Bridges: State of Science. New York

United Nations (2022) https://unsdg.un.org/2030-agenda/universal-values/leave-no-one -behind attended 2022-12-2

Wallis M (2005) Caring and evidence-based practice: The human side of critical care nursing. *Intensive and Critical Care Nursing*, 21(5): 265–267. http://hdl.handle.net/10072/7821

Wade DT (2020) What is rehabilitation? An empirical investigation leading to an evidence-based description. *Clinical Rehabilitation*, 34(11). doi:10.1177/0269215520905112

WHO (2021) https://www.who.int/news-room/fact-sheets/detail/rehabilitation attended 2022-12-2

Wilkin K & Slevin E (2004) The meaning of caring to nurses: An investigation into the nature of caring work in an intensive care unit. *Journal of Clinical Nursing*, 13(1): 50–59. doi:10.1111/j.1365-2702.2004.00814.x

Zahavi D (2014) Phenomenoligy of empathy. Chapter 10, In: *Self and Other: Exploring Subjectivity, Empathy and Shame*. Oxford University Press, Oxford

13 Technology in the Home
An Ethical Discussion about Aging Adults with Cognitive Changes

Alexandra Laghezza

INTRODUCTION

The home is an important, meaningful place in the lives of aging adults. Older adults are attached to the structural, physical components of the home and have emotional attachments they feel toward their home (Blunt & Dowling, 2006). As a result of these attachments to their homes, older adults express a strong desire to remain in their home and age in place, rather than to move to an institutional setting as they grow older. Aging in place is the desire to remain in one's home instead of moving to another setting where care is provided for them, such as a nursing home. People have expressed feelings of loss of autonomy and of being imprisoned while living in institutional care settings (Hennelly et al., 2021). In contrast, the home provides autonomy and freedom that an institutional setting does not offer. The ability to remain in one's home is dependent upon a variety of factors and not all adults are able to age in place. For example, there is a high frequency of people with dementia and Alzheimer's disease in institutional settings (Harris-Kojetin et al., 2019). This indicates that this population faces additional challenges to being able to age in place.

Older adults with age-related cognitive changes (ARCC)[1] continue to have emotional attachment to their homes (Frank, 2005. However, there are many difficulties that people with ARCC (and their caregivers) need to overcome to be able to remain in their homes safely. Some of the difficulties include caregiver stress, medicine management, an inability to complete activities of daily living (ADLs) and safety-related problems (Thoma-Lürken et al., 2018). Without effective intervention, these challenges can result in the inability of an adult to age in place. Most commonly, strategies to promote aging in place focus on home modifications to address safety-related concerns (Struckmeyer & Pickens, 2016), specifically, reducing the risk of falls in the home (van Hoof et al., 2010), such as by adding grab bars to prevent falls (Lach & Chang, 2007). Unfortunately, home modifications to support the cognitive needs of a person are not made at the same rate that modifications are made to compensate for the physical changes that occur (Marquardt et al., 2011). In other words,

 DOI: 10.1201/9781003272786-13

changes to the home are often completed to support the physical changes of the older adult, whereas not as many changes are made to the home to support the cognition changes that are experienced. The reason for this could be a lack of education and understanding about the ways that families can support an older adult with changing cognition. There continues to be a need for interventions to support the daily activities of older adults with ARCC, increase safety, and decrease caregiver burden. These interventions must support the changes, both physical and cognitive, that older adults with ARCC experience. Technological innovations are being developed and refined to support the lives of older adults aging in place.

There are wide-ranging technological innovations attempting to promote aging in place among older adults with ARCC. This field of study, also referred to as geron-technology, explores the relationship between new technological advancements and aging (Rodeschini, 2011). This technology includes sensors and GPS tracking devices to more advanced smart home technologies like integrated monitoring systems. The overall aim of the technology is to support an adult's safety in the home and increase independence. Although there are known supports that can be put in place to support this population's life at home, such as a home health aide providing assistance, home modifications, therapy interventions, etc., there continues to be an exploration of new innovations that can be placed in the home to support aging in place. Researchers have turned to technology to monitor behavior in the home and create a safer environment. However, these technologies, as they continue to be developed and refined, require an ethical framework and an interrogation of its uses and implications.

When technology is being embedded in the home to surveil and monitor the movements and behaviors in someone's personal space, there are many ethical boundaries that are at risk of being crossed. One of the many ethical concerns, specifically related to this population, centers around who is making the decision to implement this technology and the understanding of who has access to this information and how it is being used. In other words, is another person deciding what technology is being installed and what information is being collected? Questions that need to be interrogated include, was the installation of this technology approved by the older adult? After installation of this technology, are there ways for the older adult to remember that there is technology monitoring them (signs, alerts, etc.)? How is the privacy of that individual not overlooked by the need to ensure that they are safe?

Older adults wish to remain in their homes because of the freedom, privacy, security, and independence that it offers; however, how does this change with the presence of cameras, sensors, and monitors in place in the home? Additionally, does this technology remove the need for human connection? In other words, does the technology decrease the potential visits that a person may receive from family, neighbors, and friends because the older adult is now always being "checked on" through technology? What role do healthcare workers play in the development and utilization of this technology? Throughout this chapter, these and additional ethical questions are explored while presenting current research behind smart home technology and other home technologies. There must be a deep interrogation of technology that presents as beneficent devices used to maintain safety in the home. This chapter aims to critique this category of technology and pose a variety of related ethical questions. In addition, the chapter offers strategies to reduce the unethical implications of the use of the technology.

SELFHOOD

Although there are many unanswered questions related to the ethical development, installation, and utilization of these technologies, the following sections seek to further explore these questions and initiate a starting point that can be utilized before this technology is brought into the home. The goal is to center the discussion around the individual with ARCC to respect their privacy, independence, and autonomy. Even though older adults with ARCC are not the only population of adults at risk for their respect, privacy, and independence to be compromised by this technology, by focusing on the older adult with the most needs, we ensure that all older adults with varying levels of cognition are accounted for and included in the discussion. This is important because if healthcare practitioners, scholars, and researchers do not begin the discussion and center the discussion around the aging adult's needs, then there is a risk that they will quickly become unethical in their respective practices.

The foundation for this discussion is selfhood and the selfhood that an individual with cognitive changes maintains as their cognition changes and declines. The understanding of selfhood within this discussion and topic is vital to understanding that the person, despite changing cognition, is still a person with an identity that persists. It creates a starting point to ensure that their needs are not overlooked. Without this understanding and continued recognition of personhood that is maintained for this population, the ethics and respect for the individual's privacy, independence, and autonomy will always be neglected.

Despite cognitive changes, the self (self[1], personal identity) remains intact throughout one's life. However, the other aspects of the self that a person holds that are socially presented and constructed can be lost, but through no fault of the individual (self[2]). The socially presented self is only lost by the ways in which others interact with and treat the individual (Sabat & Harré, 1992). As people, we are one person but have multiple personae we present to the world (we have only one self[1], but many self[2] that are constructed based on the contexts we live in). Therefore, the social world we interact in is highly dependent on the formation and continuation of self[2] (Sabat & Harré, 1992).

The application of the theoretical concept of selfhood lends itself to the ethics of smart home technology because of the focus on the self that remains rather than focusing on the incorrect assumption, that cognitive changes results in a loss of self. With this understanding of selfhood, there is the acceptance that self[1] remains intact throughout life. When healthcare practitioners maintain this affirmation, they are less likely to neglect the needs of the individual because there will be respect for the person and their persistent self[1], and they will work to honor the individual's personal identity. An understanding and acceptance of selfhood also acknowledges the importance of the social world that people exist in and how the social world and social interactions can create a misunderstanding of behaviors. A misunderstanding of behavior without proper acknowledgments and adjustments leads to the individual being neglected. This affirmation positions the need for an appropriate social environment for the self[2] to be exhibited and understood with changes in cognition. Healthcare providers need to understand how to do this and have the necessary training to do this successfully. Healthcare providers can consistently affirm and respect the selfhood of the individual with cognitive changes and work to understand points of confusion to provide the technological intervention that is most effective and ethical. The following sections will consistently revisit the concept of selfhood as a way to explore the ethical questions and further support the utility of selfhood in ethical development, implementation, and utilization of technology in the home.

ETHICS OF TECHNOLOGY IN THE HOME: AN OVERVIEW

Ethical considerations of technology in the home include concrete concepts of boundaries, such as invasion of privacy (Morris et al., 2013), to more philosophical concerns, such as the unexplored relationship between technology and the lived environment (Baldwin, 2005). Explicit ethical considerations are not always present in the research performed in this area (Ienca et al., 2018). Our current systems of promoting the safety and security of vulnerable individuals are not ideal; nor for that matter are the new technologies. Healthcare workers, when recommending these systems, are faced with a choice between imperfect systems. Because of this, they must be very diligent in their choice and decide which system is optimal for human freedom and respects their patient's dignity (Welsh et al., 2003). This requires reflecting on a variety of different ethical questions.

When a person and their movements are monitored, who will have access to this information? Will technology reduce or completely replace human contact (Baldwin, 2005; Chung et al., 2016)? Additionally, how does reduction of human contact impact the progression of cognitive changes? Some researchers urged people to understand that gerontechnologies are not neutral safety measures but are "spinoffs" of current surveillance systems (Kenner, 2008). These surveillance systems used in dementia care are intended to control the aging body (Kenner, 2008). There is a fine line between protecting the safety of the older adult and controlling their behaviors. Although the advent of these technologies is intended to focus on safety, because of the information that can be collected, they can quickly turn into controlling devices used to monitor and control the movements of the older adult.

Unfortunately, despite the above-mentioned questions related to the ethics of gerontechnology, research studies dedicated to these concerns are few (Meiland et al., 2017). One framework used to guide research in this area had humanistic concerns at its core, including respect for the person, their abilities, and their privacy, where the research needs come second to the humanistic concerns of the individual (Mahoney et al., 2007). Whenever technology is added to a person's life, the ethical implications of that technology should be explored and extensively considered; more research is needed. Continued research is needed on the lived experience of smart care systems and these emerging technologies, in addition to clinical research (Hine et al., 2022), and how this technology can be provided in a humanizing way (Zhu et al., 2021). Based on the ethical concerns related to the use and implementation of this technology, an ethical framework is necessary. Additionally, an increase in ethical considerations could potentially result in greater use of these technologies with older adults with ARCC (Ienca et al., 2018), especially because one of the barriers to the adoption of this technology is the concern older adults have for an invasion of privacy (Morris et al., 2013).

GPS TRACKING

GPS tracking devices are used to increase safety in this population of older adults. GPS tracking can be used to track outdoor or indoor mobility and is used in the home and in institutional settings, like nursing homes. Overall, acceptance of this type of technology is high among the person using it and their caregiver (Liu et al., 2017). One reason for the high acceptance of GPS tracking is because, while using GPS tracking devices, people experienced more freedom outside and caregivers were less worried about their loved one's outdoor mobility (Pot et al., 2012). It does not restrict the movements of the user and is small enough that it is inconspicuous when worn. Specifically, caregivers greatly appreciate tracking devices (Rialle et al., 2008). One possible explanation for this finding is that the use of a GPS gives a caregiver "peace of mind" (Liu et al., 2017, p. 112). This device primarily provides a sense of security for caregivers, in case their family member wanders off, and provides additional safety measures (Landau et al., 2010). However, one of the considerations that should be taken into account when using GPS tracking is how to balance the safety of the person with how to protect their privacy. The person with ARCC should be aware of being tracked with GPS and be part of the discussion to choose to utilize it. If the individual is unable to make an informed decision, their proxy needs to be acting with the values and wishes of the person with ARCC (Landau & Werner, 2012). Despite GPS tracking's high acceptance from the older adult and the caregiver, it only addresses one concern that caregivers express relative to their loved one, namely wandering away. There are many other safety-related concerns and additional barriers that older adults experience living at home that technology seeks to solve.

SENSORS, CAMERAS, AND SMART HOME TECHNOLOGY

Smart home technology is a broad term that includes a variety of technologies to monitor and provide safety throughout one's home. This technology includes sensors, automation and remote access, and transmission of data and information to other devices outside of the home (Arthanat et al., 2019). Smart homes are designed to promote the independence and safety of older adults through monitoring technologies (Ding et al., 2011). The main objective of smart home technology is to monitor and collect data regarding a person's movements in their home to ensure safety and detect any changes in behaviors and health. However, the effectiveness of smart home technology to promote aging in place is yet to be determined (Moyle et al., 2021).

With the documented difficulties that older adults with ARCC experience in maintaining their ability to complete ADLs, technology has been developed to support the completion of daily self-care tasks and increase the likelihood of aging in place (Bossen et al., 2015; Mihailidis et al., 2004; Mynatt et al., 2000). There are vison-based systems that are being developed that would intelligently aid a person's ADL performance by automated prompting the person by "unobtrusively monitor[ing]" their behaviors (Mihailidis et al., 2004, p. 238). In other words, this system would give cues to the person during a personal care task. Similar technologies include both visual and audio prompts to promote ADL performance (Phua et al., 2009). This technology would reduce/eliminate the need for human prompting and replace it with technology to provide the needed verbal prompts that some people with ARCC require to successfully complete self-care tasks. Additional research has begun to develop and explore the use of assistive robotics in the home to provide step-by-step guidance during ADLs (Begum et al., 2013). Improved ADL performance has been a focus of technology research because there is significant evidence that shows a positive correlation between unmet ADL needs and nursing home placement. This correlation has been argued to be a better predictor of nursing home placement than measures of cognitive and functional impairment (Gaugler et al., 2005).

There are many different technologies that can be placed in the home to monitor one's movements that are not integrated like the systems in smart homes, including single cameras, light sensors, and alarms that are cheaper. These technologies are not integrated as part of an entire system, but are singular pieces of technology. There are systems that do not just focus on safety and completion of ADLs; some sensor systems focus on health events and the prediction of adverse health events in the home, such as falls (Rantz et al., 2008). Sensors are being used to monitor movement patterns and behavior with the aim of detecting any aberrant behavior (Lotfi et al., 2012). Sensors have been, and continue to be, developed to detect agitation and aggression and, ideally, work to detect patterns of behavior so that appropriate interventions can be made (Khan et al., 2018).

With the extensive amount of information that can be collected through technology, one area of concern that relates to the implementation of the technology is the lack of awareness of the technology that the older adults with ARCC may have. Are they aware that they are being monitored and watched? What can be done to increase the likelihood that the older adult will remain aware of the technology that is in the home? There should be signs placed in the home to indicate to the adult that there is technology in the home and what information is being collected. Additionally, the number of people who have access to this information should be limited to, as well as possible, protect the individual.

SOCIAL AND OCCUPATIONAL ROLES IN THE HOME

Engagement in occupations and social relationships increases the quality of life and overall well-being for older adults with ARCC (Stav et al., 2012; Öhman & Nygård, 2005). Maintaining selfhood, additionally, is achieved through social and occupational roles and keeping the aging adults concerned connected to other people

(Hennelly et al., 2021). Although occupational roles decline as an adult with cognitive changes ages, the cessation of different occupations is not always because of a change in memory or functional abilities. One possible explanation is that other people in their lives are completing the occupation for them (Lee et al., 2006). Despite the difficulty that this population faces with maintaining social and occupational roles, if the environment is supportive, these roles can be maintained. Caregivers and healthcare providers need to pay careful attention when adding technology into the home to make sure that there can continue to be an engagement in occupations in the home.

People's social environments impact their experiences, well-being, and meaning (Sabat, 1998; Sabat & Harré, 1994). What aspects of the social environment are at risk of declining, or being replaced altogether, as technology is being embedded into the home? Are we increasing the risk of isolation that is already common throughout the aging population? These questions are important because a significant portion of older adults are at a high risk of social isolation and the addition of technology should not further increase this risk of social isolation (Cudjoe et al., 2020).

The above-mentioned questions require additional research to explore these areas and find a balance within the complex environment of the home. Although the home is the primary focus for the older adult living in it, caregivers also play a vital role in the lives of the older adult. Older adults, as their cognition declines, require an extensive amount of care (Kasper et al., 2015). Oftentimes, the inclusion of technology is not only for the safety of the older adult but also to reduce the burden on the caregiver. There must be a balance that is maintained for all parties involved while ensuring that the decisions being made are ethical and based upon respect, trust, and communication.

CREATING A CONVERSATION

When technology is placed in someone's home, the question that should always be asked is, has there been a comprehensive discussion, in terms that the older adult can understand, to install this technology? Reviewing Kitwood's (1997) work indicated that, in dementia care, the relationships must be based on values of trust, respect, and communication. These values should be consistent in technologies in the home for the environments not to be oppressive. One way to create an environment that is not oppressive is to create a conversation between the older adults with ARCC and their caregiver so that the respect for the older adult is maintained. There should be a conversation between the older adult and the family/caregiver. When indicators of changes in cognition occur, these discussions should be initiated, and healthcare workers can initiate this discussion. Even if there is not a need for technological intervention, these discussions should be encouraged so that the desires of the older adult are expressed and can be honored as much as possible. It is important to introduce and discuss any technology when cognitive changes are initially noted and expected to deteriorate (Malmgren Fänge et al., 2020). Additionally, there should be shared decision-making when technology is being considered to be brought into the home. Families require support to be able to do this successfully (Berridge & Wetle,

2020). Healthcare providers can provide strategies to enhance the conversation, honoring both sides of those in the discussion. Additionally, the healthcare provider can provide education regarding the range of technology to those informed decisions which can be made that respects the privacy of the older adult and their safety needs.

In conversations between an older adult with ARCC and the caregiver, questions should be asked that map out what the older adults' desired living situation is as they age. If the older adult desires to age in their home, ask them why they want to stay in their home. The questions should elicit the older adult's connection to the space and what that space provides for them. This information is meaningful because technology in the home can impact the environment. If an older adult reports that they like the privacy of their home, placing a camera in the home can change that and an older adult needs to be aware of this potential change. However, this may not be the case if it allows them to remain in their home for longer periods of time. If healthcare practitioners are going to recommend technologies in the home there needs to be proper education regarding the technology, that includes the adult with ARCC and the caregiver/loved one.

If an older adult with ARCC cannot communicate their desires and wishes and participate in the discussion so that their desires can be expressed, their previous wishes should be honored. The decisions that are made should reflect the individual whose decision is being made for them. "Respect for previously stated wishes is a very important value that must be balanced with a concern for current burdens and benefits" (Post, 2000, p. 130). Although this author was referring to end-of-life wishes, the same must be considered for smart home technologies and all aspects of care for this population.

The home is an important space and monitoring technologies that encroach on the privacy of the older adult should be taken seriously. This can be difficult because carers value the safety and protection of their loved above possible issues of privacy invasion (White et al., 2010). Oftentimes, caregivers need to convince the older adult with ARCC to implement technology that invades their privacy but they are more likely to be convinced to use technology that supports their leisure activities (Sriram et al., 2019). Healthcare providers may be tasked with finding the balance between all the needs not only of the older adult but of their family.

CONCLUSION

There continues to be a need for research that focuses on the ethical implications and ways to maintain the respect for the person and their privacy. Unfortunetly, there is often a trade-off to be made between safety, trust, and privacy when smart home technology is installed. Research related to the development and implementation of smart home technology needs to conceptualize trust and privacy so that it can be measured instead of discussed as a general concept (Schomakers et al., 2021). Additionally, there can be more efforts placed in co-designing smart home technology, where the individual and their caregiver are actively participating in the design of smart home technology (Bourazeri & Stumf, 2018).

This aging adult population experiences a loss of self-esteem and self-identity, with increasing feelings of powerlessness as they live through cognitive changes and receive a medical diagnosis like dementia or Alzheimer's disease (Rostad et al., 2013). Throughout a time that can already feel chaotic and lonely for a person, additional measures should be in place to respect them and their ability to make choices. The literature that focuses on this technology consistently articulates the continued need for a strong ethical guideline to the research, implementation, and use of this technology. There is a consensus that more work needs to be done regarding the ethical implications of this new and fast-developing technology. Due to the lack of evidence and accepted, consistent, ethical standards surrounding this topic, the rest of this chapter highlights necessary elements that need to be accepted in advance when recommending, developing, or interacting with smart home technology for this population. At the core of the discussion is the older adult with ARCC and the self-hood that is maintained throughout changing cognitive status. Without this concept being accepted as true and factual, one cannot ensure that the technology that is being implemented will be utilized ethically.

NOTE

1. The use of the term ARCC includes medically diagnosed conditions (dementia, Alzheimer's disease) and older adults with cognitive changes that impact their daily life.

REFERENCES

Arthanat, S., Wilcox, J., & Macuch, M. (2019). Profiles and predictors of smart home technology adoption by older adults. *OTJR: Occupation, Participation and Health*, 39(4), 247-256. doi: 10.1177/1539449218813906

Baldwin, C. (2005). Technology, dementia and ethics: Rethinking the issues. *Disability Studies Quarterly*, 25(3). https://doi.org/10.18061/dsq.v25i3.583

Begum, M., Wang, R., Huq, R., & Mihailidis, A. (2013). Performance of daily activities by older adults with dementia: The role of an assistive robot. *13th International Conference on Rehabilitation Robotics (ICORR)*, 1–8. https://doi.org/10.1109/ICORR.2013.6650405

Berridge, C., & Wetle, T. F. (2020). Why older adults and their children disagree about in-home surveillance technology, sensors, and tracking. *The Gerontologist*, 60(5), 926–934.

Blunt, A. & Dowling, R. (2006). *Home*. London: Routledge.

Bossen, A. L., Kim, H., Williams, K. N., Steinhoff, A. E., & Strieker, M. (2015). Emerging roles for telemedicine and smart technologies in dementia care. *Smart Homecare Technology and Telehealth*, 3, 49–57. https://doi.org/10.2147/SHTT.S59500

Bourazeri, A., & Stumpf, S. (2018). Codesigning smart home technology with people with dementia or Parkinson's disease. In: *NordiCHI '18 Proceedings of the 10th Nordic Conference on Human-Computer Interaction* (pp. 609–621). New York: ACM. ISBN 9781450364379, ORCID: 0000-0001-6482-1973

Chung, J., Demiris, G., & Thompson, H. J. (2016). Ethical considerations regarding the use of smart home technologies for older adults: An integrative review. *Annual Review of Nursing Research*, 34(1), 155–181. https://doi.org/10.1891/0739-6686.34.155

Cudjoe, T. K. M., Roth, D. L., Szanton, S. L., Wolff, J. L., Boyd, C. M., & Thorpe, R. J. (2020). The epidemiology of social isolation: National health and aging trends study. *The Journal of Gerontology, 75*(1), 107–113. https://doi.org/10.1093/geronb/gby037

Ding, D., Cooper, R. A., Pasquina, P. F., & Fici-Pasquina, L. (2011). Sensor technology for smart homes. *Maturitas, 69*(2), 131–136. https://doi.org/10.1016/j.maturitas.2011.03.016

Frank, J. (2005). Semiotic use of the word "home" among people with Alzheimer's disease: A plea for selfhood? In G. D. Rowles & H. Chaudhury (Eds.), *Home and Identity in Late Life: International Perspectives* (pp. 171–197). New York, NY: Springer.

Gaugler, J. E., Kane, R. L., Kane, R. A., & Newcomer, R. (2005). Unmet care needs and key outcomes in dementia. *Journal of the American Geriatrics Society, 53*(12), 2098–2105. https://doi.org/10.1111/j.1532-5415.2005.00495.x

Harris-Kojetin, L., Sengupta, M., Lendon, J., Rome, V., & Valverde, R. (2019). *Long-term Care Providers and Services Users in the Unites States, 2015–2016 (No. 2019-1427)*. Hyattsville, MD: National Center for Health Statistics.

Hennelly, N., Cooney, A., Houghton, C., & O'Shea, E. (2021). Personhood and dementia care: A qualitative synthesis of the perspectives of people with dementia. *The Gerontologist, 61*(3), e85–e100. https://doi.org/10.1093/geront/gnz159

Hine, C., Nilforooshan, R., & Barnaghi, P. (2022) Ethical considerations in design and implementation of home-based smart care for dementia. *Nursing Ethics*, 1–12. https://doi.1177/09697330211062980

Ienca, M., Wangmo, T., Jotterand, F., Kressig, R. W., & Elger, B. (2018). Ethical design of intelligent assistive technologies for dementia: A descriptive review. *Science and Engineering Ethics, 24*(4), 1035–1055. https://doi.org/10.1007/s11948-017-9976-1

Kasper, J. D., Freedman, V. A., Spillman, B. C., & Wolff, J. L. (2015). The disproportionate impact of dementia on family and unpaid caregiving to older adults. *Health Affairs, 34*(10), 1642–1649. https://doi.org/10.1377/hlthaff.2015.0536

Kenner, A. M. (2008). Securing the elderly body: Dementia, surveillance, and the politics of "aging in place." *Surveillance & Society, 5*(3). https://doi.org/10.24908/ss.v5i3.3423

Khan, S. S., Ye, B., Taati, B., & Mhailidis, A. (2018). Detecting agitation and aggression in people with dementia using sensors- A systematic review. *Alzheimer's & Dementia, 14*, 824–832.

Kitwood, T. (1997). *Dementia Reconsidered*. Maidenhead, Berkshire: Open University Press.

Lach, H. W., & Chang, Y.-P. (2007). Caregiver perspectives on safety in home dementia care. *Western Journal of Nursing Research, 29*(8), 993–1014. https://doi.org/10.1177/0193945907303098

Landau, R., Auslander, G. K., Werner, S., Shoval, N., & Heinik, J. (2010). Families' and professional caregivers' views of using advanced technology to track people with dementia. *Qualitative Health Research, 20*(3), 409–419.

Landau, R., & Werner, S. (2012). Ethical aspects of using GPS for tracking people with dementia: Recommendations for practice. *International Psychogeriatrics, 24*(3), 358–366. https://doi.org/10.1017/S1041610211001888

Lee, M., Madden, V., Mason, K., Rice, S., Wyburd, J., & Hobson, S. (2006). Occupational engagement and adaptation in adults with dementia: A preliminary investigation. *Physical & Occupational Therapy in Geriatrics, 25*(1), 63–81. https://doi.org/10.1080/J148v25n01_05

Liu, L., Miguel Cruz, A., Ruptash, T., Barnard, S., & Juzwishin, D. (2017). Acceptance of global positioning system (GPS) technology among dementia clients and family caregivers. *Journal of Technology in Human Services, 35*(2), 99–119. https://doi.org/10.1080/15228835.2016.1266724

Lotfi, A., Langensiepen, C., Mahmoud, S. M., & Akhlaghinia, M. J. (2012). Smart homes for the elderly dementia sufferers: Identification and prediction of abnormal behaviour. *Journal of Ambient Intelligence and Humanized Computing, 3*(3), 205–218. https://doi .org/10.1007/s12652-010-0043-x

Mahoney, D. F., Purtilo, R. B., Webbe, F. M., Alwan, M., Bharucha, A. J., Adlam, T. D., Jimison, H. B., Turner, B., & Becker, S. A. (2007). In-home monitoring of persons with dementia: Ethical guidelines for technology research and development. *Alzheimer's & Dementia, 3*(3), 217–226. https://doi.org/10.1016/j.jalz.2007.04.388

Malmgren Fänge, A., Carlsson, G., Chiatti, C., & Lethin, C. (2020). Using sensor-based technology for safety and independence–the experiences of people with dementia and their families. *Scandinavian Journal of Caring Sciences, 34*(3), 648–657.

Marquardt, G., Johnston, D., Black, B. S., Morrison, A., Rosenblatt, A., Lyketsos, C. G., & Samus, Q. M. (2011). A descriptive study of home modifications for people with dementia and barriers to implementation. *Journal of Housing for the Elderly, 25*(3), 258–273. https://doi.org/10.1080/02763893.2011.595612

Meiland, F., Innes, A., Mountain, G., Robinson, L., van der Roest, H., García-Casal, J. A., . . . Franco-Martin, M. (2017). Technologies to support community-dwelling persons with dementia: A position paper on issues regarding development, usability, effectiveness and cost-effectiveness, deployment, and ethics. *JMIR Rehabilitation and Assistive Technologies, 4*(1), e1. https://doi.org/10.2196/rehab.6376

Mihailidis, A., Carmichael, B., & Boger, J. (2004). The use of computer vision in an intelligent environment to support aging-in-place, safety, and independence in the home. *IEEE Transactions on Information Technology in Biomedicine, 8*(3), 238–247. https:// doi.org/10.1109/TITB.2004.834386

Morris, M. E., Adair, B., Miller, K., Ozanne, E., Hansen, R., Pearce, A. J., Santamaria, N., Viegas, L., Long, M., & Said, C. M. (2013). Smart-home technologies to assist older people to live well at home. *Journal of Aging Science, 1*(1). https://doi.org/10.4172/2329 -8847.1000101

Moyle, W., Murfield, J., & Lion, K. (2021). The effectiveness of smart home technologies to support the health outcomes of community-dwelling older adults living with dementia: A scoping review. *International Journal of Medical Informatics, 153*, 10453. https://doi .org/10.1016/j.ijmedinf.2021.104513

Mynatt, E. D., Essa, I., & Rogers, W. (2000). Increasing the opportunities for aging in place. *Proceedings on the 2000 Conference on Universal Usability*, 65–71. https://doi.org/10 .1145/355460.355475

Öhman, A., & Nygård, L. (2005). Meanings and motives for engagement in self-chosen daily life occupations among individuals with Alzheimer's disease. *OTJR: Occupation, Participation and Health, 25*(3), 89–97. https://doi.org/10.1177 /153944920502500302

Phua, C., Foo, V. S., Biswas, J., Tolstikov, A., Aung, A., Maniyeri, J., Huang, W., That, M., Xu, D., & Chu, A. K. (2009). 2-layer erroneous-plan recognition for dementia patients in smart homes. *2009 11th International Conference on E-Health Networking, Applications and Services (Healthcom)*, 21–28. https://doi.org/10.1109/HEALTH.2009 .5406183

Post, S. G. (2000). *The Moral Challenges of Alzheimer Disease: Ethical Issues from Diagnosis to Dying* (2nd ed.). Baltimore,MD: The Johns Hopkins University Press.

Pot, A. M., Willemse, B. M., & Horjus, S. (2012). A pilot study on the use of tracking technology: Feasibility, acceptability, and benefits for people in early stages of dementia and their informal caregivers. *Aging & Mental Health, 16*(1), 127–134. https://doi.org/10 .1080/13607863.2011.596810

Rantz, M., Skubic, M., Miller, S., & Krampe, J. (2008). Using technology to enhance aging in place. In S. Helal, S. Mitra, J. Wong, C. K. Chang, & M. Mokhtari (Eds.), *Smart Homes and Health Telematics* (pp. 169–176). Berlin Heidelberg: Springer.

Rialle, V., Ollivet, C., Guigui, C., & Hervé, C. (2008). What do family caregivers of Alzheimer's disease patients desire in smart home technologies? Contrasted results of a wide survey. *Methods of Information in Medicine, 47*(1), 63–69. https://doi.org/10 .3414ME9102

Rodeschini, G. (2011). Gerotechnology: A new kind of care for aging? An analysis of the relationship between older people and technology. *Nursing & Health Sciences, 13*(4), 521–528. https://doi.org/10.1111/j.1442-2018.2011.00634.x

Rostad, D., Hellzen, O., & Enmarker, I. (2013). The meaning of being young with dementia and living at home. *Nursing Reports, 3*(1), 12–17. https://doi.org/10.4081/nursrep.2013.e3

Sabat, S. R. (1998). Voices of Alzheimer's disease sufferers: A call for treatment based on personhood. *The Journal of Clinical Ethics, 9*(1), 35–48.

Sabat, S. R. & Harré, R. (1992). The construction and deconstruction of self in Alzheimer's disease. *Ageing and Society, 12*, 443–461. https://doi:10.1017/S0144686X00005262

Sabat, S. R., & Harré, R. (1994). The Alzheimer's disease sufferer as a semiotics subject. *Philosophy, Psychiatry, & Psychology, 1*(3), 145–160.

Schomakers, E. M., Biermann, H., & Ziefle, M. (2021). Users' preferences for smart home automation–investigating aspects of privacy and trust. *Telematics and Informatics, 64*, 101689. https://doi.org/10.1016/j.tele.2021.101689

Sriram, V., Jenkinson, C., & Peters, M. (2019). Informal carers' experience of assistive technology use in dementia care at home: A systematic review. *BMC Geriatrics, 19*(160). 1–25. https://doi.org/10.1186/s12877-019-1169-0

Stav, W. B., Hallenen, T., Lane, J., & Arbesman, M. (2012). Systematic review of occupational engagement and health outcomes among community-dwelling older adults. *The American Journal of Occupational Therapy, 66*(3), 301–310. https://doi.org/10.5014/ ajot.2012.003707

Struckmeyer, L. R., & Pickens, N. D. (2016). Home modifications for people with Alzheimer's disease: A scoping review. *The American Journal of Occupational Therapy, 70*(1), 1–9. https://doi.org/10.5014/ajot.2015.016089

Thoma-Lürken, T., Bleijlevens, M. H. C., Lexis, M. A. S., de Witte, L. P., & Hamers, J. P. H. (2018). Facilitating aging in place: A qualitative study of practical problems preventing people with dementia from living at home. *Geriatric Nursing, 39*(1), 29–38. https://doi .org/10.1016/j.gerinurse.2017.05.003

van Hoof, J., Kort, H. S. M., van Waarde, H., & Blom, M. M. (2010). Environmental interventions and the design of homes for older adults with dementia: An overview. *American Journal of Alzheimer's Disease & Other Dementias, 25*(3), 202–232. https://doi.org/10 .1177/1533317509358885

Welsh, S., Hassiotis, A., O'mahoney, G., & Deahl, M. (2003). Big brother is watching you–the ethical implications of electronic surveillance measures in the elderly with dementia and in adults with learning difficulties. *Aging & Mental Health, 7*(5), 372–375. https:// doi.org/10.1080/1360786031000150658

White, E. B., Montgomery, P., & McShane, R. (2010). Electronic tracking for people with dementia who get lost outside the home: A study of the experience of familial carers. *British Journal of Occupational Therapy, 73*(4), 152–159. https://doi.org/10.4276/030 802210X12706313443901

Zhu, B., Shi, K., Yangs, C., Niu, Y., Zeng, Y., Zhang, N., Liu, T., & Chu, C. H. (2021). Ethical issues of smart home-based elderly care: A scoping review. *Journal of Nursing Management*, 1–14. https://doi.org/10.1111/jonm.13521

14 Determining the Role of Socially Assistive Robots in Healthcare

Chris McCarthy, Sonja Pedell, Felip Martí, Jo Butchart, Joel D'Rosario, Sarah Knight, and Adam Scheinberg

SOCIALLY ASSISTIVE ROBOTS

Socially assistive robots (SARs) aim to effect therapeutic benefit through social interaction and companionship (Feil-Seifer & Matari, 2005). Over the past 20 years, SARs have emerged as a promising technology, demonstrating encouraging results across a range of health and well-being use cases, including aged care (Wada et al., 2005), autism spectrum disorder (ASD; Boccanfuso et al., 2016), play therapy (Bers et al., 1998), and physical rehabilitation (Wade et al., 2011), to name but a few. Proposed benefits from the use of SARs in healthcare settings range from more effective/efficient care delivery through enhanced compliance with prescribed exercises such as in pediatric rehabilitation, distraction/entertainment when experiencing pain/discomfort, better emotional and mental health outcomes through social engagement and companionship (e.g., oncology (Alemi et al., 2016), stroke rehab (Feingold-Polak et al., 2021)), enhanced communication (ASD, brain injury) (Ilyas et al., 2019), to efficiency gains in care delivery through reduced stress on the workforce (e.g., robot triaging (Chang & Murphy, 2007)).

Previous research has proposed and, to varying levels, demonstrated the potential benefits of SARs in a range of healthcare settings. However, much of this work has focused on proof-of-concept outcomes or the demonstration of specific technical capabilities in lab-based studies. As the number of field-tested SARs steadily grows, so too does the urgency for more focus on the alignment of SAR design and development goals with the targeted clinical and well-being outcomes, and the needs for ongoing deployment in healthcare settings such as hospitals, clinics, and residential care facilities. In this regard, there has been only limited attention paid to the design, deployment, and evaluation of SARs in real settings, integrated as part of existing care delivery systems.

DOI: 10.1201/9781003272786-14

SOCIALLY ASSISTIVE ROBOT RESEARCH

Feil-Seifer and Matarić's seminal paper, "Defining socially assistive robotics" (Feil-Seifer & Matarić, 2005) provided the first definition of SARs as a distinct research field. In this work, they assert SARs as robots that aim to effect therapeutic benefit through social interaction, thereby distinguishing them from more traditional applications of robots focused on physical assistance. Much of SAR research derives from the broader field of human–robot interaction (HRI), encompassing both physical interactions, focused on manipulation and haptics, and social-emotional aspects of robots interacting with humans (Matarić & Scassellati, 2016). Research and evidence reporting on SARs has steadily progressed since the turn of the century, with increasing contributions from other related fields such as human–computer interaction, behavioral psychology (e.g., Martin et al., 2020), and design.

Indeed, SAR research has experienced a substantial increase in interest over the past ten years. Figure 14.1, from a literature search on SCOPUS, plots the number of peer-reviewed journal articles published over the past 20 years. This increased focus on social robots can be attributed to a number of factors, including advances in AI, reductions in hardware costs making robotics and mechatronics research more accessible, and, perhaps most significantly, the emergence of affordable commercially available social robots offering off-the-shelf platforms to develop and explore SAR use cases. The availability of general-purpose social robots spanning humanoid (or human-like) systems, such as Softbank Robotics' Nao and Pepper robots, as well as more pet-like systems, such as the Sony Aibo and PARO, the robot seal, to more recent systems such as MiRo-E, among many others, have vastly improved the accessibility of robot platforms, as well as their underlying capabilities through

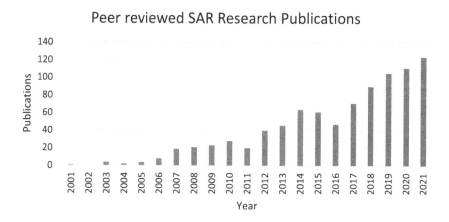

FIGURE 14.1 SAR-related peer-reviewed publication counts over the past 20 years. Numbers are obtained from SCOPUS search on the terms "therapeutic robot" OR "socially assistive robot," included in the title, abstract, or keywords. While not exhaustive, this graph is indicative of the rapid rise in interest in SARs, particularly over the past five years.

the deployment of increasingly high-level software development interfaces, enabling easy scripting and prototyping of SAR behaviors and activity scenarios.

SARs in the Field: Roles, Design, and Assessment

Social robot research has extensively explored populations and use cases in which SARs may offer benefits. These populations include, but are not limited to, specific age groups, impairments, and vocations. Feil-Seifer and Matarić (2005) defined a set of population groups within which some benefit can be gained through a socially assistive robot. Specifically, they highlighted the following groups:

1) Elderly
2) Individuals with physical impairments
3) Individuals in convalescent care
4) Individuals with cognitive disorders
5) Students

In Table 14.1, we summarize example studies from the literature spanning such population groups, along with details of the size and duration of each study with respect to validation of the SAR with each targeted population.

These examples highlight a diversity of roles for which SARs have been considered. In some cases, the choice of role is firmly dictated by the targeted outcome; for example, SARs designed to address loneliness or social interaction typically take on roles of companionship, e.g., Tapus et al. (2009) and Wada et al. (2002). In other settings, however, choices of role exist which must be carefully considered and often combined. In contexts such as ASD, for example, studies such as Clabaugh et al. (2019) combine companionship and demonstration, addressing two distinct needs for that population. Other SAR studies considering ASD, as well as in other child counseling contexts have shown benefits when deploying a SAR as a mediator, offering a more comfortable channel of communication between child and adult care provider or educator (Pennisi et al., 2016). The coaching role is most commonly associated with motivation, such as is often employed during physical rehabilitation (e.g., in pediatric rehabilitation (Pulido et al., 2019) or post-rehabilitation (Matarić et al., 2007)), bringing with it an element of instruction and the expectation of feedback.

This diversity of roles available reflects the unique versatility that SARs offer for healthcare applications. However, while priority use cases are well established, the field remains largely in an exploratory phase with regard to SAR roles, with more focus required on long-term evaluations. There is a growing acknowledgment of the need for evaluations of SARs for specific targeted outcomes within the context of their intended use (Martí et al., 2018). However, undertaking extended SAR trials in complex contexts with clinical populations and vulnerable groups in need of care necessitates the development of SARs beyond proof of concept, and the adoption of design and evaluation processes that can support this.

In the next sections, two case studies from different health domains are presented, examining roles for SARs in different therapeutic settings, namely pediatric

TABLE 14.1

An Overview of Field Trial Studies of SARs Summarizing Various Use Cases and Roles Assigned to SARs, and the Extent of Testing and Evaluation Performed

Theme/group	Roles	Location	Sample	Outcome	Reference
Interaction with the elderly	Companion, mediator, entertainer	Japan, 2002	N = 26, 6 weeks	+ Decreased stress in the elderly + Decreased burnout of the caregivers	Wada et al. (2002)
Cognitive therapy with dementia patients	Motivator, facilitator	California, USA, 2009	N = 4, 6 months	+ Improved performance in game + Improved reaction time	Tapus et al. (2009)
Aiding motion-impaired people	Human proxy	Stockholm, Sweden, 2002	N = 1, 3 months	+ Seen as useful or necessary − May be a distraction for bystanders	Huttenrauch & Eklundh (2002)
Coaching post-stroke patients	Motivator	California, USA, 2007	N=6, 6 sessions each	+ Higher attendance *versus* no robot	Matarić et al. (2007)
Coaching cerebral palsy patients	Motivator, demonstrator	Sevilla, Spain, 2019	N = 8, 4 months	+ Improved mobility	Pulido et al. (2019)
At-home personalization with children with autism	Companion, motivator, demonstrator	California, USA, 2019	N=17, >30 days each	+ Generally engaged with participants + Improved math scores	Clabaugh et al. (2019)
Remote teaching of students	Human proxy, demonstrator	Japan, 2011	N = 22, 1 session	+ Increased motivation towards class− Can be seen as scary	Hashimoto et al. (2011)

rehabilitation and aged-care (dementia respite care) settings. These contexts represent different use cases with specific requirements and targeted outcomes, while also highlighting commonalities in the underlying design challenges to overcome, limitations of current SAR technology, and practical considerations that accompany the integration of SARs for ongoing use in such healthcare settings.

CASE STUDY 1: A SAR FOR PEDIATRIC CASE STUDY

PROJECT SETTING

The Royal Children's Hospital, Melbourne, Australia hosts a busy pediatric rehabilitation clinic. A particularly prominent patient group is those children with cerebral palsy (CP). For many children living with CP, orthopedic surgery is required to correct secondary musculoskeletal problems which impact on gait and function during early childhood. Such patients typically undergo up to three rehabilitation sessions per day, over a two- to three-week period as in-patients at the hospital (Thomason et al., 2013). Although therapists supervise a majority of these rehabilitation sessions, approximately one-third of sessions are supervised by an adult parent/guardian or on-ward nurse – typically with the guidance of a pre-prepared hardcopy printout describing exercises to perform and the number of sets and repetitions for each exercise. It is challenging to motivate young patients to complete multiple daily sessions of rehabilitation while also regulating their mood and ensuring exercises are performed correctly. Hence sessions without therapist supervision are less likely to be completed as prescribed, or correctly executed, and thus are generally assumed to contribute less to the overall rehabilitation success of the patient.

The augmentation of rehabilitation care delivery through the use of a SAR was thus proposed to promote increased motivation, and compliance in patients during rehabilitation sessions – in particular sessions without therapist supervision. Specifically, software for the humanoid Nao social robot (Softbank Robotics) was developed to adapt the Nao robot for use as a therapeutic aid for rehabilitation. The SAR development and evaluation spanned a total duration of approximately four years, spanning three phases of development: Phase One: Exploration; Phase Two: Iterative development and evaluation; and Phase Three: Integration for unsupervised on-ward single-patient multi-session care delivery. The entire study was approved by The Royal Children's Hospital Melbourne Human Research Ethics Committee (HREC: 36128C) and Swinburne's Human Research Ethics Committee (SUHREC: 2016/202).

PHASE ONE: EXPLORATION

A regular weekly pattern of visits to the clinic was established in the early weeks of this phase (McCarthy et al., 2015). Each week, the Nao robot was set up in the clinic waiting room, close to consultation rooms, ensuring high visibility to

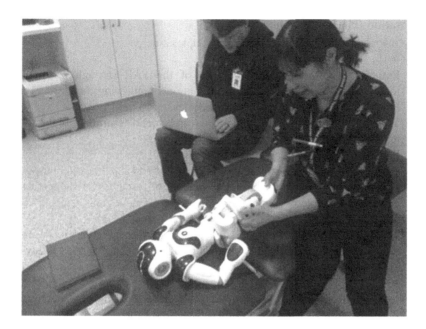

FIGURE 14.2   Phase One exploration included inviting therapists and carers to physically interact and manipulate the robot to explore its range of motion, physical capabilities, and limitations.

patients, their families, and therapists. This facilitated regular, albeit brief, discussions with therapists and parents at the beginning. Patient interactions were initially also brief, unstructured, and intermittent, occurring before and between therapy sessions. Wizard-of-Oz control was predominantly used to invoke and facilitate interactions.

Early engagement suggested how to overcome the technology limits and foster effective engagement with patients. It facilitated the development of core exercise demonstrations. Therapists were actively engaged in this process, initially through requests to critique Nao's execution of exercises, and also invited to physically manipulate the robot's limbs to both correct and explore the physical capabilities and limitations of the system (Figure 14.2).

In the second half of this phase, therapist engagement evolved into a cycle of iterative development in which a therapist directly programmed specific exercises by positioning the robot into key poses, from which robot joint positions were immediately recorded and time-sequenced. New exercises were rapidly developed via this process on-site, with refinements made between clinic visits.

Patient engagement also progressed from non-specific patient interactions driven primarily by general interest in the robot, to the inclusion of Nao in therapist-selected patient sessions. Pre-built exercise demonstrations were sequenced in accordance with therapist specifications and trialed in sessions with technical support.

Exploration (Phase One) Outcomes

Phase one identified five key stakeholder groups, namely:

- **Patients:** The primary beneficiaries of the SAR through potentially increased motivation and sustained emotional well-being, faster recovery time; and improved rehabilitation outcomes.
- **Therapists**: Primary users of the system and chief determinants of the SAR therapeutic assistance role and fitness for purpose.
- **Parents/Guardians/Carers**: Primary duty of care for patients, targeted end-users, and determinants of the system's usability and fitness for purpose.
- **Hospital/Clinic Administrators**: Providers of resources to support care delivery, with interests in high-quality, cost-effective, and scalable care delivery.
- **Technology Developers** gather feedback from other stakeholders, develop the system, and assess the system's technical performance and feasibility of identified roles.

Stakeholder engagement also determined four specific SAR roles encompassing base-level capabilities deemed important for the SAR to serve as an effective therapeutic aide:

- **Demonstrator:** SAR performs the exercise in front of the child and provides verbal instruction to emphasize important aspects of the exercise.
- **Motivator**: SAR provides verbal encouragement throughout the session, before and during each prescribed exercise, along with enticements (e.g., music, dancing, and joke telling) to encourage exercise completion.
- **Companion**: SAR delivers a personalized introduction to build rapport and perform each exercise with the child, delivering encouraging observational statements.
- **Coach**: SAR guides the patient through the session by scheduling, coordinating, and pacing the execution of the above roles to deliver a complete session of therapy.

Finally, to support the above roles, Phase One identified the following system requirements.

- **Configurability**: therapists requested a system that allows pre-selection of exercises to perform, the number of repetitions, speed of execution, and entertainment modules, as well as personalization of the session with the patient.
- **Stability**: SAR actions must operate with a high degree of certainty in order to minimize session interruption and distraction. SAR movement, in particular, must be carefully designed to operate within the working limits of the system.

- **Adaptability**: SAR should provide mechanisms for dynamic adjustment of activity settings, including number of repetitions, speed, and sequence order. Verbal instructions must adjust accordingly.
- **Interaction**: SAR interaction should be multimodal (e.g., verbal, tactile, etc.) to cater for varying patient needs as well as system speech recognition limitations.
- **Integration**: SAR must be easily set up, portable, and transportable by a single person, and operable by carers with minimal training requirements.
- **Responsiveness**: SAR should exhibit sufficient responsiveness to unprompted verbal statements from patients to maximize the perceived authenticity of the SAR role as a companion.
- **Stand-alone**: SAR should be operable without engineering support, Wizard-of-Oz control, or additional hardware to meet the needs of flexible and un-hindered ongoing use.
- **Robustness and Endurance**: SAR should operate continuously and for a minimum of 30 minutes without engineer intervention.

PHASE TWO: ITERATIVE DEVELOPMENT AND EVALUATION

Phase Two prioritized the *in-situ* iterative development and evaluation of a stand-alone SAR prototype based on Phase One outcomes and observations (Martí Carrillo et al., 2018). Phase Two thus focused on iteratively developing a system capable of leading rehabilitation sessions.

The SAR Prototype

Based on Phase One observations, a software platform was developed. The SAR was equipped to guide patients through approximately 30-minute rehabilitation sessions. Sessions consisted of multiple exercises, each involving several sets and repetitions. For each exercise, the SAR presented a demonstration while explaining key features of the exercise. The patient was then invited to join the SAR in completing a set together. The Phase Two prototype was equipped with 13 different rehabilitation exercises: a sit-to-stand exercise (Figure 14.3a, left image) and 12 exercises performed from a lying down position (nine examples shown in Figure 14.4). The SAR was also equipped to run activities in which the child was guided to perform a task by the robot. Figure 14.3b (right) depicts an activity scenario in which the robot guides patients through a so-called toy relay game. In this scenario, the robot asked the patient to fetch named toys on the other side of the room. During execution of the exercise, the SAR provided encouragement and therapist-selected reminders about key aspects of each exercise. Adjustments to exercise speed could be made by the carer using a tactile interface on the robot's head via simple button taps. Upon completion of a session and throughout, the SAR offered rewards through joke-telling and pre-scripted dance routines.

FIGURE 14.3   Upright exercises and activities. 14.3a Left: Sit-to-stands in which the SAR asks the child to stand up and touch its head, and then sit down. 14.3b Right: "Toy-relay," in which the SAR instructs the child to walk across the room and collect a toy and bring it back.

Iterative Development and Evaluation

Patients, parents, and therapists were all formally recruited to participate in testing, as part of an *in-situ* iterative development process. The inclusion criteria for patients were that they had been prescribed a rehabilitation program consistent with the SAR predominantly lower-limb exercise capabilities (due to focus on cerebral palsy rehabilitation needs) and based on a physiotherapist's clinical judgment. Informed consent was obtained from all parents (on behalf of themselves and their child) and physiotherapists participating in the study. After completing a session, participants were given the option to participate in another session. If agreed to, the attending parent was invited to operate the robot themselves without a physiotherapist present.

Figure 14.5 shows images from the testing environment, including where sessions with the SAR were undertaken and the adjacent observation room where researchers observed each session.

Data Collection

Eliciting user perceptions of the system was a targeted outcome of the phase, with a focus on physiotherapist, parent, and patient responses to the SAR design, and their experience with it during care delivery. Sessions were observed by research team members in an adjacent room (Figure 14.5b (right)). At the end of each session, patients, parents, and therapists completed two surveys: an adapted version of the "Acceptance of an assistive social robot" questionnaire, developed by Heerink et al. (2009) (e.g., Table 14.2) shows adapted questions for therapists), and the GodSpeed Survey (Bartneck et al., 2009). Tailored, open-ended questions were also asked of all participants at the completion of each session.

PHASE TWO OUTCOMES

Child Perceptions

A total of 20 patients (nine female, 11 male) interacted with the SAR across 41 sessions. Patients ranged in age from five to 16 years old. Survey responses after each

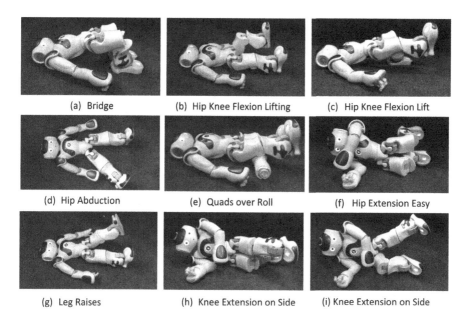

(a) Bridge   (b) Hip Knee Flexion Lifting   (c) Hip Knee Flexion Lift

(d) Hip Abduction   (e) Quads over Roll   (f) Hip Extension Easy

(g) Leg Raises   (h) Knee Extension on Side   (i) Knee Extension on Side

FIGURE 14.4 Examples of rehabilitation exercises implemented for the Nao and used in Phase Two trials. All are commonly prescribed for patients with cerebral palsy undergoing rehabilitation after musculoskeletal corrective surgery.

FIGURE 14.5 Images of the testing environment used for Phase Two trials showing (left, 14.5a) the consultation room where the patient, therapist and parent(s) worked with the SAR during each session, and (right, 14.5b) the adjacent observation room with one-way mirror used by researchers to observe sessions.

session generally reflected positive attitudes towards the SARs use in their sessions, with most patients reporting that they enjoyed working with the robot. More than two-thirds of patient responses indicated that the robot was useful in their rehabilitation session, with a further 21% neutral.

Post-trial interviews highlighted more polarized attitudes and perceptions of SAR use – in particular for older children (11 years+). One older child described the robot as "patronizing" and another felt annoyed as the robot was not listening to her. In

TABLE 14.2

Adapted Acceptance Questionnaire of Heerink et al. (2009) Used for SAR User Feedback from Therapists during Phase Two of the SAR Development

Construct	No.	Question
ANX1	1	I would be afraid to make mistakes using the robot
	2	I would be afraid to break something when using the robot
ANX2	3	I find the robot scary
	4	I find the robot intimidating
ATT	5	it's a good idea to use the robot
	6	The robot would make therapy sessions more interesting
FC	7	I have everything I need to make good use of the robot
	8	I know enough of the robot to make good use of it
ITU	9	If I have access to the robot, I think I'll use it during the next therapy sessions
	10	If I have access to the robot, I am certain to use it in the next therapy sessions
	11	If I have access to the robot, I'm planning to use it during the next therapy sessions
PAD	12	I think the robot can be adaptive to what I need
	13	I think the robot will only do what I need at that particular moment
	14	I think the robot will help me when I consider it to be necessary
PEOU	15	I think I will know quickly how to use the robot
	16	I find the robot easy to use
	17	I think I will be able to use the robot without any help if I have been trained
	18	I think I will be able to use the robot when there is someone around to help me
	19	I think I will be able to use the robot when I have a good manual
PU	20	I think the robot is useful to help in pediatric therapy
	21	It would be convenient to have the robot for therapy sessions with kids
	22	I think the robot can help me with many things during pediatric sessions
SI	23	I think the staff would like me using the robot
	24	I think parents would like me using the robot
	25	I think patients would like me using the robot
	26	I think it would give a good impression if I should use the robot
TR	27	I would trust the robot if it gave me advice
	28	I would follow the advice the robot gives me

general, older children very much considered the SAR as a robot/computer rather than a companion. However, no child expressed a desire for the robot to be less machine-like but rather for it to interact with more depth and complexity, and to be more responsive.

Parent Perceptions

Parents were the participant group who was the most positive about the use of the SAR in their child's care. Parents identified their child's affective experience during rehabilitation sessions as being highly important and valued that the SAR was able to improve their child's engagement during sessions. Parents also identified multiple benefits of the SAR facilitating increased independence with exercises, allowing some ability to hand "responsibility" over to the SAR, thereby taking pressure off the parents' own relationship with the child. The SAR robot became responsible for demonstrating, instructing, and talking the child through the exercises, reducing both the time and emotional burden on the parent. Parents, who operated the SAR themselves in a subsequent session (after only observing in the first session), were universally more positive in their responses to the SAR's usefulness and its usability compared to the first session responses, indicating that experience with the SAR helped greatly to shape positive attitudes toward it.

Therapist Perceptions

Eight physiotherapists operated the robot to deliver rehabilitation sessions across Phase Two testing. Acceptance surveying and open-ended responses indicated physiotherapists' attitudes toward using the SAR in rehabilitation sessions were universally positive. Therapists also overwhelmingly agreed that the SAR was useful in their sessions and was easy to use. Notably, therapists also indicated that parents and patients felt positive about their use of the robot in their rehabilitation sessions.

Therapists identified that greater adaptability in the system was desirable, in particular with respect to patients' performance within the session. Therapists also felt the SAR should be able to monitor the child's performance and provide specific feedback, in addition to the general feedback and encouragements provided by the current system. Therapists highlighted that system or hardware failures were disruptive and diminished trust in the system's reliability, and their willingness to keep using it.

PHASE THREE: ONWARD INTEGRATION

Phase Three evaluated the performance of the SAR prototype with respect to the primary use case initially defined – to be fully integrated into the clinical program of selected patients, and in daily use. This phase took the form of focused case studies, removing all technical support during sessions. The aim was thus to assess the integration of SARs with existing clinical care, and to elicit any further requirements that emerged when used in this fully integrated context.

Configurability via Tablet Interface

Phase Two highlighted the critical importance of configurability both prior and during sessions with the SAR. The Phase Three prototype was thus equipped with a tablet interface, co-designed with therapists by a team of software engineering students. The tablet interface provided therapists with a tool to configure and start the robot program themselves. In addition, the tablet offered the ability to set the pace of exercise demonstrations, and include other interactive modules like games and dances, as well as add critical information, such as the patient's name, and what specific session the configuration was for. Sessions could be saved and recalled for specific patients.

In addition to pre-session configuration, the tablet offered dynamic control of the SAR, allowing the operator to adjust the schedule during sessions, and adjust the pace of exercises based on their observations in the session. Figure 14.6 shows a sample screenshot from the tablet interface.

Phase Three Trial Setting and Procedure

Phase Three testing ran for approximately one week per family, with the nominated parent running daily rehabilitation sessions consecutively. For each session, the SAR was delivered to the patient's room by a staff member, with the rehabilitation program already configured into the system via the tablet interface. Parents were also

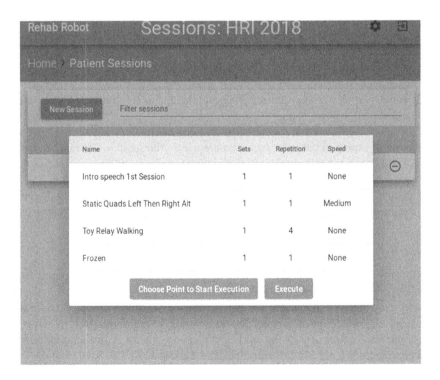

FIGURE 14.6 A sample screenshot of the tablet interface used by therapists and parents to configure and control the SAR.

provided with a small survey to self-report outcomes after each session. Patients did the rehabilitation exercises on their beds with the robot placed on the bed tray table. Figure 14.7a (left) shows the typical location of the SAR when used in the case studies, and Figure 14.7b (right), depicts the trolley used for easy transport within the hospital.

During observed sessions, the Objective Structured Assessment (OSA) form derived from the physiotherapist's interviews was used to evaluate, via observations, the quality of care delivered using the SAR under the parent's guidance. In addition, a self-report survey was filled out by the attending parent after every session.

Phase Three Evaluation Metrics

An adapted OSA form, commonly used to assess the quality of therapist-delivered sessions, was developed specifically to evaluate the success of the SAR in supporting the delivery of rehabilitation therapy without direct physiotherapist supervision. To derive appropriate metrics for the OSA, physiotherapists who participated in Phase Two were recruited to participate in a follow-up study consisting of individual semi-structured interviews (Butchart et al., 2019).

A thematic analysis on the interview transcripts was then performed, from which five themes were generated for the evaluation of the success of a session being delivered by a parent to their child, summarized in Table 14.3.

For each identified theme, a checklist of observable items was generated, from which a form was produced for evaluating the quality of pediatric rehabilitation care delivery.

Phase Three Outcome Summary

Three families (four patients, including one family with twins) were recruited to undertake the trial. Each family delivered rehabilitation independently for three or four days (after the early sessions used for training), and the last session was

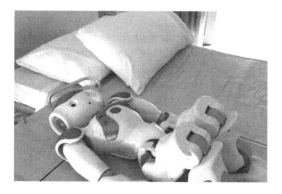

FIGURE 14.7   The SAR on-ward in Phase Three. The left image (14.7a) shows the Nao robot on the tray table where it performed most activities. The right image (14.7b) shows the SAR transportation and storage trolley.

TABLE 14.3

Summary of Themes Extracted from Therapist Interviews and Derived Observable Indicators for Assessing Sessions with SAR in Phase Three

Themes	Key Performance Indicators	Derived Observable Indicators for Evaluating SAR-Assisted Care Delivery
Technique	Ensuring correct movement, posture/position, and speed	Extent to which robot exercise demonstrations are utilized to explain exercise and/or a cause of distraction
Carer understanding	Demonstrating an understanding of purpose and aim of each exercise. Retention from initial therapist-led run-through	Extent to which the SAR interactions, capabilities, and configurable settings supported: 1. prompting of next exercise/activity to undertake 2. carer's explanation of what exercise/activity to undertake 3. session moderation to adjust to patient's needs
Carer's engagement	Extent of active engagement in completing the rehabilitation session correctly	Extent to which the carer is observing and helping child while working with SAR to deliver care
Patient engagement	How actively engaged patient is in completing exercises	Willingness of patient to interact with the SAR and extent to which patient is engaged with and positive about completing the exercises in accordance with SAR guidance
Robot presence	Extent to which SAR offers observable benefit to the carer-patient dynamic, *versus* effort to set up and operate	Observed use of SAR to successfully invoke and complete exercise/activity. Perceived effort to use SAR. Recorded SAR system failures/issues

observed by the research team. In general, parents showed competence and confidence when operating the robotic system, and most did not express any concerns or anxieties when using it independently. Based on observations using the Objective Structure Assessment form, parents were successfully able to deliver rehabilitation with the help of the SAR on the ward.

Patients' completion of the exercises varied depending on the patient, their behavior, and whether they got distracted or disengaged. In general, all patients had a positive attitude toward the robot throughout the week, although some observations indicated waning interest in the robot by the last session for one patient.

The focus on daily operation of the SAR in the intended place of its use highlighted a range of insights into the integration of the SAR within existing practice that was not evident from the weekly Phase Two testing alone. Overall, Phase Three findings were summarized as follows:

- **SAR Roles** The relative importance of some roles, deemed important from Phase One and Phase Two, changed when using the robot in daily sessions. In particular, the robot's demonstration of the exercises became a source of distraction for patients already familiar with the exercise, suggesting options to skip this should be incorporated.
- **Personalization** Patients valued the SAR's ability to refer to them by name, although, in some cases, they expressed frustration at the SAR's lack of responsiveness or awareness of their current state. This highlighted a clear barrier in the SAR's current ability to build and maintain rapport over time, which was clearly something valued by some of the patients in the study.
- **Setup** Families were responsible for booting up the SAR once delivered to the location. The Nao robot's relatively prolonged bootup sequence (about five minutes) was observed to introduce challenges in maintaining the child's positive emotional state. Notably, in Phase Two, this was never an issue as the SAR was in a ready state from the moment the child entered the consultation room.
- **SAR Stability** As in Phase Two, the SAR was subject to loss of stability at times, and this was observed in self-reporting to occasionally induce worry in the patient for the wellbeing of the SAR. This was a notable difference to Phase Two, where technical and/or therapist staff were on hand to quickly intervene. However, the SAR failure recovery programming, introduced for Phase Three, was able to successfully rectify the situation with the assistance of the parent.
- **Stand-alone, Ready-to-use SARs** Despite some noted challenges above, the SAR's design as a stand-alone therapeutic aid requiring minimal setup (beyond positioning) and pre-configured program of care delivery successfully enabled the completion of every session in all Phase Three case studies.
- **Configurability** Configurability of the system was observed to be substantially better than in Phase Two and central to the SAR's successful deployment in Phase Three evaluations. This was primarily attributable to the tablet interface through which parents were able to start the SAR with the corresponding session programmed and the pace settings adjusted. Further configurability could easily be included in the tested system, and, in general, as many options as possible to adapt sessions to patient needs should be supported in any SAR designed for clinical deployment.
- **Environment** Interruptions on the ward were frequent and unavoidable, so SARs should be designed with this in mind. Simple examples include providing easy ways to pause the rehabilitation session in execution. Advances in artificial intelligence may also detect such events, and self-pause, or ask for confirmation to continue.

CASE STUDY 2: A SAR FOR INCREASING WELL-BEING OF PEOPLE LIVING WITH DEMENTIA

PROJECT SETTING

The global cost of dementia in 2022 was expected to reach US$ 1.3 (Alzheimer's Disease International, 2019). There is a global estimate of 82.1 billion hours for 2015 or an equivalent of 5.7 hours per day of informal care provided per person living with dementia (Wimo et al., 2018). Hence, there is a need for support for informal carers through respite and new and different forms of stimulating activities in such settings (Roger et al., 2012). The research team investigated the potential benefits of modern technologies in such settings for care groups of older people living with dementia and reports here specifically on the use of Nao robots in a council's active aging program with a focus on people living with dementia. Socially assistive robots and their acceptability and success have been investigated in residential care settings (Khosla et al., 2017) and their ability to support the process of care giving to increase the well-being of older adults is promising (Kachouie et al., 2014). However, beneficial scenarios and goals around the use of Nao robots in group activities with people living with dementia are scarce and poorly defined.

It can be a challenge to keep people, living with dementia, occupied with engaging activities due to their often-short attention span and declining cognitive abilities. Humanoid robots provide quite a sophisticated presence which was expected to be helpful to increase engagement by this participant group. Another advantage of the Nao robot is that there is no need to consider input devices which can be problematic due to limitations of older users' abilities, such as sensory and motor impairments or simply a hesitation to touch unfamiliar technology. Despite these limitations, older people have diverse needs and interests depending on their life experiences and circumstances. Meaningful activities that enhance those people's well-being and quality of life are crucial. This case study reports on the support a Nao robot can provide to staff with the care of people living with dementia in activity groups.

Meaningful activities in dementia care groups that enhance the lives of people living with dementia include a focus on mobility and music. Robots have been used to deliver interventions to prevent physical decline for the elderly as such activities can be programmed and help to retain independence (Robinson et al., 2014). In this regard, the robot becomes a preventative health tool. Hence, a combination of music- and mobility-related activities was considered promising at the beginning of this research. The project received ethics approval of the Swinburne University Human Research Ethics Committee (SUHREC #2012/305) and older adult participants, their guardians, and staff gave informed consent before the research commenced.

According to the review of Robinson et al. (2014), most robots have been used to address individual needs or mediate one-to-one interactions. However, a one-carer-to-one-dementia-user model is not feasible for providing care in most aged care settings. In our scenario, social group activities with a robot are expected to be valuable, where a single caregiver is supported by a robot to engage a group of people with dementia in an activity. The PARO seal, for example, has been successfully used to spark conversations among older adults who normally would not talk to each other (Shibata, 2012).

This research investigated the integration of the humanoid Nao robot into a care group setting of fourteen older adults living with dementia through iterative visits in the community hall over several months. The research team regarded a group setting to be the ideal environment where the control remains with the individual, whether they want to interact with the robot or not. However, the complex group setting of dementia care also requires that one cannot merely rely on the functionality of the robot, but it needs a holistic approach to plan and evaluate interactions with robots, which also include perspectives on social structures and everyday context. Acknowledging that the older adults and staff are the experts of their life context and needs, the team decided to follow an approach of developing the human–robot interactions and evaluating these, guided by the principles of co-design (Brandt et al., 2012).

PHASE ONE: EXPLORATION

During every visit, the values of co-design were applied, and attention paid to the context of the group and its individuals. A mixed-method approach, consisting of interviews, observations, researcher notes, video analysis, and interaction studies, was applied to evaluate the level of engagement and how the groups reacted to and interacted with the robot. After the visits, the team members discussed their observations and notes in debriefing sessions and defined goals for the next visit based on the results of the current visit. This is comparable with the procedure of reflexive practice described by Taylor et al. (2018). In Phase One, the group activities offered by the staff were investigated for their suitability to include the robot. The objective was to choose the most promising activities to integrate the robot and design and later evaluate the interactions with the robot *in situ* with the older adults and staff.

Key Stakeholder Groups

When the team brought the robot to the active aging group for people living with dementia for the first time, "it" was met with a lot of skepticism. Group members had no idea what a robot could do, and many were uncomfortable and had a negative perception of robots in general. Also, the staff, which was one of the key stakeholder groups, were fearful that "robots might one day replace them" or "take over the world." The research team learned that, to be beneficial – and with a broader aim to increase well-being – the robot needed to integrate into the daily group activities and create some real purpose without taking away from staff, and taking these fears seriously. In addition, based on these reactions, the team decided to reduce the abstract and technical appearance of the Nao robot. The robot was named Kira during the study and the researchers also asked the council's knitting group whether they could help create some clothes for the Nao (Figure 14.8 images on the right). These factors contributed to the robot's identity making it easier to integrate Kira into the group, to approach Kira, referring to "her," and acknowledge her as a stakeholder in her own regard.

FIGURE 14.8 Testing of the stability of the robot during movement, using different objects to sit on, and after receiving some clothes.

Role of the Robot

After observing the range of group activities offered by staff, the research team settled on the exercise and dancing activity for Kira to be implemented. The reasons were that these activities were always attended by the whole group, very popular, and, according to staff, had the biggest effect on their well-being in terms of creating a positive mood and calming some of the more agitated group members. Investigating the robot's potential roles (e.g., investigator or group member), it was decided for the robot to become a co-instructor alongside the staff member. The researchers could not risk the robot instructing incorrect movements hence increasing the risk of injury for the older adults. It was also anticipated that familiar staff leading the instructions would maintain the momentum, as the robot was not able to do all the required movements (e.g., tilting neck to shoulders), could not move in an obvious manner due to its physical structure (e.g., rotating shoulders), and its small size made it more difficult to discern movements from afar. In implementing the robot as a staff assistant, the group members had the choice and control to follow either the robot or the instructor.

Key Requirements

To prepare the exercise sessions, an instruction manual with descriptions and photos of a staff member conducting the exercises was borrowed. All exercises were conducted in a sitting position to also include group members with mobility issues. A total of 22 exercises were programmed which followed the weekly group activity. The exercise part was followed by some dancing. This part included songs that are supported by sing-along and dance moves. The popular Australian song "Home among the Gum Trees" was chosen which has a set of moves accompanying the lyrics which are familiar to people throughout Australia. Apart from the exercises, this song was performed by a staff member for the research team to videorecord the matching moves and song lyrics. The song was performed by Kira stranding up – the group members would either sit or stand depending on their mobility and balance.

Interaction with the Group

It needed to be planned how to integrate the robot to support the interactions of the group. As the exercises contributing to the largest part of the session were carried out by the group while sitting, the robot was also seated but on a table for better visibility. Maintaining stability in the sitting position was a challenge, as the Nao robot maintains balance best when standing or squatting. Numbers of constructed chairs, mini sitting balls, and stools were trialed by the research team for best balance and optimal height (see Figure 14.8). The sitting position was specifically programmed by the developer, accommodating the stool that turned out to be most suitable – a kid's step stool with a rubbery surface.

Interaction Mechanisms

The team had to decide not only **what** activities to use the robot for, but also **how** the robot was to interact with the group. There are several technical interaction mechanisms Nao robots have available. These include voice recognition, tactile sensors, infrared sensors, or control through a computer. It was decided that the movements and an introductory speech would be programmed, and the timing would be controlled from a laptop. Voice control can cause possible delays when the robot does not recognize the commands. Tapping on the tactile head sensors to move through the list of exercises can also be slow. Computer control turned out to be the least interruptive interaction mechanism as Kira could be controlled from the background without visible input, giving the impression that she was autonomous (Figure 14.9, image on the right). It was also the quickest way to flexibly jump to exercises and start new subprograms. The biggest challenge in the preparation was to plan a continuous sequence of movements and anticipate the right timing in tune with the staff member as the original programming was done from the paper manual.

PHASE TWO: ITERATIVE DEVELOPMENT AND EVALUATION

Physical Exercise Program with the Group

Successful interactions during the physical exercise routine included looking at Kira and copying her movements. When the group realized Kira was also doing the exercises, they shifted their attention from Sally (staff member) to Kira, with some of them listening to Sally, but watching Kira. Some group members copied the speed at which the robot demonstrated the exercise, but, when Kira was too fast, Sally had to intervene. Timing was the biggest challenge during the exercise part. Although Sally

FIGURE 14.9   Kira doing stretching exercises with the group (hip stretch and neck stretch).

tried to match her movements to Kira's, she gave preference to the timing requirements for the group to have the exercise done "properly." Overall, the staff member
supported the robot integration by directly addressing Kira: "How are you going,
Kira?" She also reinforced the role of Kira as an instructor saying: "Now, we do one
round on one foot only, like Kira." When the instructing staff would do an exercise
that the engineer had not programmed, she would say something like: "Kira can't do
this one as she does not have five fingers, but we are doing it [opening and closing
a fist]." However, the incapability or limits to movements did not seem to worry the
group members at all.

Interestingly, there was a sense of pride among the older adults when they were
doing things better than Kira. Sympathy was shown toward Kira when she wobbled
or trying to keep balance, as all participants were conscious of their own physical
limitations. This engendered more interest from the group.

Movements Kira did particularly well were often complimented with positive
responses. When the movements were correct and consistently timed, they were still
perceived as robot-like in comparison to Sally. This often made the group, and in
particular the staff, laugh. The sections where staff, older adults, and robots were in
complete sync created a feeling of achievement among the group. From the observations, it is fair to conclude that the range of capabilities, in particular the limitations,
made her more human. Engagement of the older adults with Kira during the exercises was evident through their attention to and interest in her. One particular success
was pointed out by an observing staff member regarding a participant who often
becomes agitated: "There is no music and Jack is sitting still – this is unheard of."

In sum, the most challenging aspect of the integration of Kira into the group
exercises was the difficulty of timing to achieve a smooth interaction. Hence, a lot
of development time was spent on the refinement of the timing of the SAR. The
most surprising insight was the high emotional engagement with Kira, in particular
the positive response to her incapability of some of the exercises. Because of these
discoveries, rather than trying to have perfect her movements as an instructor, the
developer worked on designing more interactions with a focus on emotions, such as
Kira making a comment during an exercise or suddenly sneeze. These readjustments
also slightly changed the perception of Kira over time in emphasizing her role as an
entertainer and group member, as well as an instructor. This also encouraged the
group joking with and talking to Kira during the exercises.

Dancing Activity

The dancing of the SAR, following the exercise activity along with the music,
worked particularly well (Figure 14.10). The song, "Home among the Gumtrees,"
was presented by Kira and the staff in the correct timing and sequence. Every group
member was involved in some way according to their capabilities – either dancing
along or singing along. If group members were unable to dance, tapping and clap-
ping were observed, demonstrating sustained engagement (Figure 14.10, left image).
Given Kira was standing during this part of the session, moving forward and backward from the hip, it needed to be ensured that the clothing was not interfering with
these movements or her balance. The standard outfit became a crocheted two-piece

FIGURE 14.10 Nao robot (Kira) dancing with staff and older adults to "Home Among the Gumtrees."

set, consisting of a short sleeve top and a knee-length skirt (see Figure 14.10, middle image).

PHASE THREE: EVALUATION, ADAPTATION IN CONTEXT AND FUTURE OPPORTUNITIES

The staff were very positive about Kira's involvement in the group activities after several visits. The main benefits of interacting with Kira they saw according to our interview results were engagement and stimulation:

The staff was surprised by the high acceptance and engagement with the robot. One staff member said: "When [Kira was] brought out, two elderly members were looking at the fish tank. Then, when there was [Kira], that was it, the fish tank was gone." Also, Kira was commented on as an asset, making the staff's life easier as some sort of novelty and stimulation tool, with opportunities to get her involved in other activities: "I think [Kira's] possibilities are endless, I really do. [...] I think you could utilize her endlessly throughout the day. I would love to have her sit down with some clients and give a history run...or even an opportunity for clients to sit down and listen to poetry." For things like that, staff would often "revert to iPads, but to have [Kira] would be even more beneficial."

Exposing the elderly to cutting-edge technology was seen as highly relevant. According to the program manager, the older participants were "intrigued to see where technology is going, particularly at their age where they haven't been introduced to technology like we have been." Consequently, group members talked about the robot afterward, which shows that they were interested and were embracing the robot. "After you left ...they would talk about [Kira]. And when they talk about her, to me, that's someone that's taken her on board. To me, that's a bonus, that's a plus."

Some group members were more engaged during exercise when the robot was present. Sally was pointing out that "When Kira is there, it seems to enthral them and they'd copy her moves, which they weren't doing before. And we've noticed in the weeks after that, that they were becoming more active," even when Kira was not participating in the session.

Staff also had suggestions for future use as robots are seen as reliable and predictable in their interactions: "I would love her to engage with the clients as they enter the facility with a greeting, "How are you?," the date, and possibly the weather [...] just a nice little welcome to both clients and carers." Speaking to the older participants, one lady stated that she wanted to take the robot home. She elaborated that she wanted to dance with Kira and liked her friendliness as "robots can't be cruel." The fact that robots are more stable in their behavior than humans, as they don't have moods, has also been discussed by Sharkey (2014) in regard to robots' qualities in the context of care of the elderly.

Benefits of Group Setting for Human–Robot Interaction

There were several benefits of the group setting for the interactions between older people with dementia and Kira. Firstly, the group setting encouraged participation as only one group member was needed to provide the catalyst for more hesitant individuals to join in. In a one-on-one setting, people would feel more pressured and probably less inclined to interact with the robot. Having the possibility to remain a member of the audience, they were still part of observing more active interactions between Kira and other group members. This enabled them to be part of the discussion and to step up and interact at their own pace.

Kira was seen by the staff as a great support within the group activities. The robot constituted a novel form of interaction stimulating group members known to have a short attention span. At times, the robot facilitated conversation between staff and the group and helped to hold or regain attention, specifically during the exercises. During some exercises, Kira became a second facilitator in her own right when she kept moving and staff demonstrated alternatives for less capable participants. Kira was well liked as it was obvious that she would not be a threat for replacing staff, but being useful in relatively brief spurts of interaction. This is important as there are serious concerns that, in care environments, robots might lead to the loss of jobs, the main rationale for SARs to be employed being to increase efficiency. In contrast, it is suggested here that, without acceptance by staff, robots cannot be established in a useful fashion.

The group setting also facilitates better use of the robot. As the Nao is quite expensive, and careful design of interactions and consequent programming of the robot is necessary, it is valuable when a larger number of people can benefit from it at the same time. From a product life-cycle point of view, it seems more realistic that councils and care homes can sustain the costs and required maintenance of robots than single households.

Designing Social Interactions with SARs

The research team set out to design beneficial social interactions with the Nao robot in an activity group for people living with dementia. Here, insights are formulated as recommendations for designing SARs in a dementia group setting based on the findings. It is suggested that these recommendations could be more generalizable to care for older adults without dementia or in different settings.

- Intensive research of the care goals and group setting is needed before the robot is deployed. This includes consulting a domain expert to discuss use scenarios before going with the robot into the field for implementation.
- Activities and length of deployment need to be chosen carefully to set realistic expectations for the target audience and engage group members in interactions that are based on their interests and not merely on their declining health or care needs.
- Choosing the role of robot carefully is crucial. The level of engagement and quality of interaction with the robot differs, largely depending on its role. The role also influences the goals for the interaction and hence how to evaluate success such as efficient care versus increased well-being.
- The more users are able to relate to the robot, the more likely it is that they engage. Display of limitations and emotions can foster this, even though they seemingly might not add to the actual care goals.
- Providing common ground and a familiar setting are crucial for the SARs in a care group with people living with dementia – not overdoing "novelty" or trying to revolutionize the setting is important. A careful balance between stimulation and familiarity needs to be maintained.
- Commitment of care staff is a precondition for the success of SARs and should be ascertained, in addition to approval by management.
- Sequence and timing are essential, and the robot programmer should have access to video footage of the intended interactions previous to the introduction of the robot. Ongoing video analysis to revise programs is also helpful.
- There should be plans for the unexpected and for alternatives. Under no circumstances should the group or individual participants feel they did not meet expectations or feel pressured to interact with the robot.

KEY CHALLENGES FOR SARS

The case studies presented describe *in-situ* design processes that emphasize stakeholder engagement, resulting in SAR systems demonstrably supporting care delivery and operating within existing care delivery systems. These case studies exemplify design processes focused on understanding context and deriving roles for SARs based on this. Both case studies also make clear current limitations in the technology but, critically, provide useful context in which to understand the relative impact and importance of specific shortcomings, and the challenges faced by current SAR research and development. We discuss the key challenges facing SAR research in more detail below.

TECHNICAL CHALLENGES FOR SARs

Natural Language Processing

Dialog-based interaction with SARs hinges critically on the underlying Natural Language Processing (NLP) competency of the robot. NLP has seen recent successful translation into commercially available smart devices, such as Apple's Siri, or Google's Alexa. However, the inability of NLP to handle non-typical voice characteristics commonly encountered, as well as only partially formed sentences, is well documented (Russo et al., 2019). Variations of human voice with respect to age, ethnicity, demographic, and gender, among an array of other influencing factors, make clear the challenges that must be overcome by any system seeking to interact naturally and in real-time.

Robot Vision

Robot vision is built on the development of computer vision algorithms, analyzing camera-captured images and/or video streams. The use cases of SARs and social robots often require visual competencies to support human-centric contexts, such as face recognition and face tracking, as well as more nuanced competencies, such as human action recognition, emotion recognition, and human intent prediction. These areas of current computer vision research are attracting significant attention, although state-of-the-art techniques still struggle to meet the needs of reliable performance in real-world settings. Additional sensing, such as wearable motion sensing or depth sensing cameras, can assist, but need additional hardware and/or calibration and setup requirements.

Machine Learning

Machine learning underpins most algorithmic advances in SAR research. Specifically, deep learning, built on the fundamentals of multi-layered deep neural networks (loosely modeling the properties of biological neural networks), has had a profound impact on AI research in general, including SAR research and development (LeCun et al., 2015). However, its impact on human–robot interaction and/or teaming has been less pronounced than in other areas of AI. A key challenge in the SAR context is supporting systems to adapt to previously unencountered scenarios, tasks, or dialogs, while also retaining competencies already gained (known as **continual learning** (Churamani et al., 2020). Coupled with this is the need for trained models capable of handling atypical input data, such as impeded speech, or human movement impacted by underlying disabilities (Hutchinson et al., 2020).

DESIGN AND INTEGRATION CHALLENGES FOR SARs

Interaction Design and Rapport Building

The survey by Leite et al. (2013) of long-term social robot studies makes clear the challenges of designing interactive behaviors that foster relationship building and sustainment (Leite et al., 2013). Routine behaviors include customary actions, such as greetings, goodbyes, etc., while strategic behaviors seek to foster and strengthen relationships through reflective communication and/or deliberate actions to foster the relationship. Leite et al. argue the latter remains a challenge for SARs. Ten years on from this review, and the challenges remain.

The case studies presented in this chapter offer insights into how appropriate SAR role formulation and choices of care delivery model can offer alternative ways to incorporate more strategic relationship building. In the rehabilitation context, the SARs' extensive configurability options with respect to dialog, entertainment, and game play options equip the SAR with strategic actions, such as acknowledging a shared history with the child and their developing relationship together, and adopting more nuanced language more catered towards the interests and personality of the child. The second case study showed promise in the relationship building when humor and emotions were involved, and similar and familiar activities were enhanced through the robot. Although not autonomously selected, such a design provides a bridge between human–robot Interaction Wizard-of-Oz studies, that have extensively explored these ideas, and full autonomy, by leveraging configurability and the SARs' pairing with a human carer. Also, in the second case study, the staff working in aged care felt supported but not replaced by the robot, which was important given that the aim was to enrich activities, rather than make them more efficient.

Form Factor and Appearance

Studies have previously highlighted the importance of a robot's embodiment and appearance (Lohse et al., 2008). From an online survey, Lohse et al. reported that more respondents thought appearance of social robots was of greater importance than its functionality. This was in the context of a domestic robot, although, in Case Study 1, simple choices, such as the color of the SAR, were observed to impact significantly on how children reacted to the system. SAR form factor in the context of aged care (and in particular dementia) represents a less-studied question, though a diversity of robots have been explored in this context. In Case Study 2, the older adults with dementia could much more easily relate to the robot once it had some knitted clothes on as this was a visual outcome they could relate to (most of them had created some clothes for a grandchild before). Even the creation of the *persona* of Kira allowed them to talk to the robot more easily. While it did not strictly change the appearance of the Nao robot, it changed how it was perceived.

Reliability

Robotic systems encompassing moving joints, advanced sensing, and complex software bring inevitable risks of failure. In the case studies presented, and in clinical settings more generally, system failures have been observed to quickly diminish trust in the system's reliability and, ultimately, it's acceptance. On-going use demands high-quality engineering and the establishment of robust failure recovery strategies. System designers must therefore give careful thought to how movements can be utilized while also ensuring that the SAR operates comfortably within its working limits. In the aged care setting, it was important that the robot should show exercises accurately and in sync with the staff to not confuse the group members. However, movements did not need to be "perfect." Being able to lift the arm higher than the robot created feelings of achievement and led to some competition, in particular, as the staff pointed out, when Kira was a bit "stiff" or not as versatile (three instead of five fingers) as the other group members. Technology not being more accurate or powerful in all aspects helped the older adults to overcome their fears and hesitation to interact.

CONSIDERATIONS FOR FUTURE SAR RESEARCH, EVALUATION, AND INTEGRATION

TRUST AND ETHICAL CONSIDERATIONS

As the use of social robots in the domestic market and within care facilities increases exponentially over the next decade (as opposed to industrial robots which are already widespread), there is a need for strong ethical frameworks and charters to guide their use. Ethical charters have been applied to human–robot interactions (HRIs) as far back as Asimov in 1942. In 2006, the European Robotics Research Network (EURON) created a Roboethics Roadmap (Veruggio, 2006). Other frameworks have been suggested, including PAPA (Veruggio, 2006) and CERNER (Grinbaum et al., 2017). Terms such as "roboethics" have been used to incorporate "ethical questions about how humans should design, deploy, and treat robots," with areas specific to HRI including discrimination of users and robots, dehumanization of users, and deception by robots (Wullenkord & Eyssel, 2020). Other sensitive areas raised by the use of SARs include their involvement in warfare, for sexual pleasure, or to care for vulnerable populations.

Key ethical considerations in relation to the use of SARs within the context of health and care industries include privacy and data protection (who owns data collected by the SAR, how will data be used, where will data be kept?), responsibility for the robot (who is responsible if it malfunctions, causes harm, or is damaged), autonomy of the SAR (how much is the SAR under the therapists' control, how much can the SAR do independently of the therapist?), potential for replacement (will the SAR reduce stress and improve the efficiency of the health care worker or will it replace the worker?), social cues (is the patient engaged, does the SAR respond with human-like gestures, like facial movements?), and trust and safety (can the SAR be trusted to do what it is meant to do and is it safe?). Etemad-Sajadi and colleagues, in a study focused on social robots in service delivery, found that social cues, trust and safety, responsibility, and privacy and data protection were the most significant factors in users' intention to use the robot (Etemad-Sajadi et al., 2022). Privacy, security, liability, and dehumanization are other concerns raised by experts in the field of HRIs in relation to trust of robots (Fosch-Villaronga et al., 2020).

Ethical considerations can be amplified when SARs are used within vulnerable populations, including children, the elderly, and those with disability, or within institutions, such as hospitals, aged care homes, or residential facilities (Sharkey & Sharkey, 2021). In these contexts, questions arise such as use of SARs as companion robots (e.g., PARO) and whether they should be used as replacements for human–human interactions.

Different cultures will view SARs through different ethical lenses. For example, in some cultures, the use of SARs in the care of the elderly or those with disability can be seen as problematic. Cultures which are more individualistic may have difficulty trusting SARs which don't function autonomously. Yet, increased SAR autonomy (moving away from the Wizard-of-Oz technique) places considerable responsibility on the programmer and the machine. One should then ask whether the SAR has been programmed for ethically compliant behavior, if there are sufficient feedback loops to avoid injury to the patient, and what is the ulterior motive of the SAR designer (Coeckelbergh, 2012)?

Human-robot Collaboration as a Path Forward

The integration of fully autonomous SARs in care-giving roles remains a long-term objective. However, SAR systems trialed to date have generally relied on human operator support either to initiate specific actions, provide dialog, assist failure recovery, or simply adjust and configure the SAR during operation. Despite this, human assistance of the SAR is rarely explicitly defined as part of the design, instead being viewed as a work-around to account for the SARs limitations or resulting system shortcomings. The recent emergence of **collaborative robots** (cobots), designed to operate safely and seamlessly with humans and in the presence of humans, offers valuable insights into how design can leverage the advantages of both human and robot competencies to increase productivity. Cobots have had most impact in manufacturing contexts, although their design and applications offer a path forward for SAR design, in particularly establishing systems around this collaborative relationship.

Establishing SAR Roles and Benefits through the ICF framework

The acceptance and integration of SARs in healthcare hinge critically on establishing therapeutic benefits in a scalable, measurable, but also meaningful way. This has proven difficult to date, with the substantive design and development complexities of SARs a key impediment to the undertaking of controlled validation studies (Robinson et al., 2019). Establishing exactly what therapeutic benefits SARs offer, and perhaps more importantly, what SARs are best suited to targeting, also remain open questions.

A framework for pursuing these questions exists through the International Classification of functioning, disability and Health ICF (2001) which offers a comprehensive and universally applicable model for understanding and measuring patient health and function.

The WHO ICF is a biopsychosocial approach to understanding functioning and disability in the presence of a health condition. The domains of body structure and function, activity, and participation are seen to dynamically interact with contextual components in the person or environment, affecting an individual's functioning. The ICF provides a common language and conceptual basis for the classification and assessment of health and disability. The framework encourages focus on the impact of interventions beyond the structure and function level to include activity- and participation-level outcomes that are often more meaningful to patients and their families. SARs represent a key environmental factor that has the potential to influence outcomes across the ICF domains.

The ICF framework incorporates functional components, but also personal, social, and environmental factors, offering potential guidance in the roles and use cases SARs may target, as well as in what and how specific benefits offered by the SAR may be measurably targeted. The ICF thus offers a much-needed, principled, and practical model for SAR development and evaluation that currently does not exist. In the context of the two case studies presented and previous work, the design processes applied could be adapted to incorporate an ICF-consistent approach.

CHAPTER SUMMARY

Socially Assistive Robots offer exciting possibilities for the augmentation of healthcare delivery to relieve growing pressure on hospitals, clinics, and residential care facilities, and meet the needs of specific patient cohorts. Advancements in artificial intelligence, sensor technologies, and computing hardware have brought SARs to a point beyond mere proof-of-concept to systems now capable of being deployed and evaluated. However, while technological advances enable this, to do so they increase the urgency for the derivation of roles for SARs that best utilize their benefits while also respecting and accommodating their limitations and operational constraints. This chapter has described examples of SAR design and evaluation processes in healthcare based on *in-situ* co-design, development, and evaluation. This resulted in the development of SAR prototypes capable of operating effectively to deliver care and achieve positive acceptance by end users. Although current limitations of the technology are also made clear in these case studies and other similar studies, understanding these shortcomings in the context of targeted outcomes, and in the context of their use, offers a way forward for prioritizing research in the field and ensuring that future advances target the need. Coupled with this is a need for well-designed and targeted measures upon which to assess therapeutic benefit. The World Health Organization (WHO) International Classification of Functioning (ICF) framework offers a holistic, evidence-based approach to achieving this, ensuring that all facets of care delivery are considered in the design and evaluation of SARs.

REFERENCES

Alemi, M., Ghanbarzadeh, A., Meghdari, A., & Moghadam, L. J. (2016). Clinical application of a humanoid robot in pediatric cancer interventions. *International Journal of Social Robotics*, 8(5), 743–759.

Bartneck, C., Kulić, D., Croft, E., & Zoghbi, S. (2009). Measurement instruments for the anthropomorphism, animacy, likeability, perceived intelligence, and perceived safety of robots. *International Journal of Social Robotics*, 1(1), 71–81. https://doi.org/10.1007/s12369-008-0001-3.

Bers, M. U., Ackermann, E., Cassell, J., Donegan, B., Gonzalez-Heydrich, J., DeMaso, D. R., Strohecker, C., Lualdi, S., Bromley, D., & Karlin, J. (1998). Interactive storytelling environments: Coping with cardiac illness at Boston's children's hospital. In *Proceedings of the SIGCHI Conference on Human Factors in Computing Systems*, 603–610. https://doi.org/10.1145/274644.274725.

Boccanfuso, L., Barney, E., Foster, C., Ahn, Y. A., Chawarska, K., Scassellati, B., & Shic, F. (2016). *Emotional Robot to Examine Different Play Patterns and Affective Responses of Children with and without ASD*, 19–26.

Brandt, E., Binder, T., & Sanders, E. B.-N. (2012). Tools and techniques: Ways to engage telling, making and enacting. In *Routledge International Handbook of Participatory Design*. New York : Routledge, 145–181.

Butchart, J., Harrison, R., Ritchie, J., Martí, F., McCarthy, C., Knight, S., & Scheinberg, A. (2019). Child and parent perceptions of acceptability and therapeutic value of a socially assistive robot used during pediatric rehabilitation. *Disability and Rehabilitation*. https://doi.org/10.1080/09638288.2019.1617357.

Chang, C., & Murphy, R. R. (2007). Towards robot-assisted mass-casualty triage. In *2007 IEEE International Conference on Networking, Sensing and Control*, 267–272.

Churamani, N., Kalkan, S., & Gunes, H. (2020). Continual learning for affective robotics: Why, what and how? In *2020 29th IEEE International Conference on Robot and Human Interactive Communication (RO-MAN)*, 425–431.

Clabaugh, C., Mahajan, K., Jain, S., Pakkar, R., Becerra, D., Shi, Z., Deng, E., Lee, R., Ragusa, G., & Matarić, M. (2019). Long-term personalization of an in-home socially assistive robot for children with autism spectrum disorders. *Frontiers in Robotics and AI, 6*, 110.

Coeckelbergh, M. (2012). Can we trust robots? *Ethics and Information Technology, 14*(1), 53–60.

Etemad-Sajadi, R., Soussan, A., & Schöpfer, T. (2022). How ethical issues raised by human–robot interaction can impact the intention to use the robot? *International Journal of Social Robotics*, 14, 1103–1115.

Feil-seifer, D., & Matari, M. J. (2005). Socially assistive robotics. *IEEE Robotics & Automation Magazine* , 18(1), 24–31.

Feingold-Polak, R., Barzel, O., & Levy-Tzedek, S. (2021). A robot goes to rehab: A novel gamified system for long-term stroke rehabilitation using a socially assistive robot methodology and usability testing. *Journal of NeuroEngineering and Rehabilitation, 18*(1), 1–18.

Fosch-Villaronga, E., Lutz, C., & Tamò-Larrieux, A. (2020). Gathering expert opinions for social robots' ethical, legal, and societal concerns: Findings from four international workshops. *International Journal of Social Robotics, 12*(2), 441–458.

Grinbaum, A., Chatila, R., Devillers, L., Ganascia, J.-G., Tessier, C., & Dauchet, M. (2017). Ethics in robotics research: CERNA mission and context. *IEEE Robotics and Automation Magazine, 24*(3), 139–145.

Hashimoto, T., Kato, N., & Kobayashi, H. (2011). Development of educational system with the android robot SAYA and evaluation. *International Journal of Advanced Robotic Systems, 8*(3), 28.

Heerink, M., Kròse, B., Evers, V., & Wielinga, B. (2009). Measuring acceptance of an assistive social robot: A suggested toolkit. In *Proceedings of the - IEEE International Workshop on Robot and Human Interactive Communication*, 528–533. https://doi.org/10.1109/ROMAN.2009.5326320.

Hutchinson, B., Prabhakaran, V., Denton, E., Webster, K., Zhong, Y., & Denuyl, S. (2020). Unintended machine learning biases as social barriers for persons with disabilitiess. *ACM SIGACCESS Accessibility and Computing, 125*, 1.

Huttenrauch, H., & Eklundh, K. S. (2002). Fetch-and-carry with CERO: Observations from a long-term user study with a service robot. In *Proceedings 11th IEEE International Workshop on Robot and Human Interactive Communication*, Berlin, Germany, 158–163.

Ilyas, C. M. A., Schmuck, V., Haque, M. A., Nasrollahi, K., Rehm, M., & Moeslund, T. B. (2019). Teaching pepper robot to recognize emotions of traumatic brain injured patients using deep neural networks. In *2019 28th IEEE International Conference on Robot and Human Interactive Communication (RO-MAN)*, 1–7, New Delhi, India.

International, A. D. (2019). *World Alzheimer Report 2019: Attitudes to Dementia*. London: Alzheimer's Disease International.

International Classification of functioning, disability and Health ICF (2001). Geneva: World Health Organization.

Kachouie, R., Sedighadeli, S., Khosla, R., & Chu, M.-T. (2014). Socially assistive robots in elderly care: A mixed-method systematic literature review. *International Journal of Human-Computer Interaction*, 30(5), 369–393.

Khosla, R., Nguyen, K., & Chu, M.-T. (2017). Human robot engagement and acceptability in residential aged care. *International Journal of Human–Computer Interaction*, 33(6), 510–522.

LeCun, Y., Bengio, Y., & Hinton, G. (2015). Deep learning. *Nature*, 521(7553), 436–444.

Leite, I., Martinho, C., & Paiva, A. (2013). Social robots for long-term interaction: A survey. *International Journal of Social Robotics*, 5(2), 291–308.

Lohse, M., Hegel, F., & Wrede, B. (2008). Domestic applications for social robots: An online survey on the influence of appearance and capabilities. *Journal of Physical Agents*, 2(2), 21–32.

Marti Carrillo, F., Butchart, J., Knight, S., Scheinberg, A., Wise, L., Sterling, L., & McCarthy, C. (2018b). Adapting a general-purpose social robot for paediatric rehabilitation through in situ design. *ACM Transactions on Human-Robot Interaction (THRI)*, 7(1), 1–30.

Martin, D. U., MacIntyre, M. I., Perry, C., Clift, G., Pedell, S., & Kaufman, J. (2020). Young children's indiscriminate helping behavior toward a humanoid robot. *Frontiers in Psychology*, 11, 239.

Matarić, M. J., Eriksson, J., Feil-Seifer, D. J., & Winstein, C. J. (2007). Socially assistive robotics for post-stroke rehabilitation. *Journal of NeuroEngineering and Rehabilitation*, 4(1), 1–9.

Matarić, M. J., & Scassellati, B. (2016). Socially assistive robotics. In Siciliano, B., & Khatib, O. (eds) *Springer Handbook of Robotics*, Cham: Springer, 1973–1994. https://doi.org /10.1007/978-3-319-32552-1_73

McCarthy, C., Butchart, J., George, M., Kerr, D., Kingsley, H., Scheinberg, A. M., & Sterling, L. (2015). Robots in rehab: Towards socially assistive robots for paediatric rehabilitation. *OzCHI 2015: Being Human - Conference Proceedings*. https://doi.org/10.1145 /2838739.2838791.

Pennisi, P., Tonacci, A., Tartarisco, G., Billeci, L., Ruta, L., Gangemi, S., & Pioggia, G. (2016). Autism and social robotics: A systematic review. *Autism Research*, 9(2), 165–183.

Pulido, J. C., Suarez-Mejias, C., Gonzalez, J. C., Ruiz, A. D., Ferri, P. F., Sahuquillo, M. E. M., de Vargas, C. E. R., Infante-Cossio, P., Calderon, C. L. P., & Fernandez, F. (2019). A socially assistive robotic platform for upper-limb rehabilitation: A longitudinal study with pediatric patients. *IEEE Robotics and Automation Magazine*, 26(2), 24–39.

Robinson, H., MacDonald, B., & Broadbent, E. (2014). The role of healthcare robots for older people at home: A review. *International Journal of Social Robotics*, 6(4), 575–591.

Robinson, N. L., Cottier, T. V., & Kavanagh, D. J. (2019). Psychosocial health interventions by social robots: Systematic review of randomized controlled trials. *Journal of Medical Internet Research*, 21(5). JMIR Publications Inc. https://doi.org/10.2196/13203.

Roger, K., Guse, L., Mordoch, E., & Osterreicher, A. (2012). Social commitment robots and dementia. *Canadian Journal on Aging/La Revue Canadienne Du Vieillissement, 31*(1), 87–94.

Russo, A., & others (2019). Dialogue systems and conversational agents for patients with dementia: The human–robot interaction. *Rejuvenation Research, 22*(2), 109–120.

Sharkey, A. (2014). Robots and human dignity: a consideration of the effects of robot care on the dignity of older people. *Ethics and Information Technology, 16*, 63–75.

Sharkey, A., & Sharkey, N. (2021). We need to talk about deception in social robotics! *Ethics and Information Technology, 23*(3), 309–316.

Shibata, T. (2012). Therapeutic seal robot as biofeedback medical device: Qualitative and quantitative evaluations of robot therapy in dementia care. *Proceedings of the IEEE, 100*(8), 2527–2538.

Tapus, A., Tapus, C., & Mataric, M. J. (2009). The use of socially assistive robots in the design of intelligent cognitive therapies for people with dementia. *IEEE International Conference on Rehabilitation Robotics, 2009*, 924–929.

Taylor, J. L., Soro, A., Roe, P., Lee Hong, A., & Brereton, M. (2018). "Debrief O'Clock" planning, recording, and making sense of a day in the field in design research. In *Proceedings of the 2018 CHI Conference on Human Factors in Computing Systems*, (CHI '18). Association for Computing Machinery, New York, NY, USA, Paper 308, 1–14. https://doi.org/10.1145/3173574.3173882

Thomason, P., Selber, P., & Graham, H. K. (2013). Single event multilevel surgery in children with bilateral spastic cerebral palsy: A 5 year prospective cohort study. *Gait and Posture, 37*(1), 23–28. https://doi.org/10.1016/j.gaitpost.2012.05.022.

Veruggio, G. (2006). The euron roboethics roadmap. In *2006 6th IEEE-RAS International Conference on Humanoid Robots*, , Genova, Italy, 612–617. doi: 10.1109/ICHR.2006.321337

Wada, K., Shibata, T., Saito, T., Sakamoto, K., & Tanie, K. (2005). Psychological and social effects of one year robot assisted activity on elderly people at a health service facility for the aged. In *Proceedings of the 2005 IEEE International Conference on Robotics and Automation*, 2785–2790. https://doi.org/10.1109/ROBOT.2005.1570535.

Wada, K., Shibata, T., Saito, T., & Tanie, K. (2002). Robot assisted activity for elderly people and nurses at a day service center. *Proceedings 2002 IEEE International Conference on Robotics and Automation (Cat. No. 02CH37292), 2*, 1416–1421.

Wade, E., Parnandi, A. R., & Matarić, M. J. (2011). *Using Socially Assistive Robotics to Augment Motor Task Performance in Individuals Post-stroke/RSJ International Conference on Intelligent Robots and Systems*. IEEE, 2403–2408. https://doi.org/10.1109/IROS.2011.6095107.

Wimo, A. W., Gauthier, S., & Prince, M. (2018). *Global Estimates of Informal Care*. London: Alzheimer's Disease International.

Wullenkord, R., & Eyssel, F. (2020). Societal and ethical issues in HRI. *Current Robotics Reports, 1*(3), 85–96.

Index

Printed in the United States
by Baker & Taylor Publisher Services